The Unofficial   *Ages 8+*

Harry Potter

Coloring Math Book

Multiplication & Division (A)
within 1000 without Regrouping

Copyright © 2019 STEM mindset, LLC. All rights reserved.

STEM mindset, LLC 1603 Capitol Ave. Suite 310 A293 Cheyenne, WY 82001 USA
www.math-stepbystep.com   info@stemmindset.com

STEM mindset™ and Brainers™ are the trademarks of STEM mindset, LLC.

This book is unofficial and unauthorized. It is not authorized, approved, licensed, or endorsed by J. K. Rowling, her publisher, or Warner Bros. Entertainment Inc, or any other person or entity owning or controlling rights in the Harry Potter name, trademark, or copyrights.

The purchase of this material entitles the buyer to reproduce worksheets and activities for classroom use only – not for commercial resale. Reproduction of these materials for an entire school or district is strictly prohibited. No part of this book may be reproduced (except as noted before), stored in retrieval system, or transmitted in any form or by any means (mechanically, electronically, photocopying, recording, etc.) without the prior written consent of STEM mindset, LLC.

First published in the USA 2018.   ISBN 9781948737531

Fun with

Contents

Multiplying and Dividing within 100 without Regrouping.....1

Multiplying by 2-9s using arrays..

Learning Multiplication Facts ....

Dividing by 2-9s ....

Learning the Order of Operations ....

Multiplication Tricks & Strategies ....

Word Problems………………………………………………..

Word Search & Mazes......………………………...…………..

Multiplying and Dividing within 1000 without Regrouping. 61

Multiplying and Dividing Tens, Hundreds, and Thousands ....

Dividing by 10-90s ....

Learning the Order of Operations ....

Multiplication and Division Tricks & Strategies ....

Word Problems………………………………………………..

Word Search & Mazes………………………………..………...

CogAT© test prep………………………………...……………..

Answers………………………………………………………111

Sources

Rowling, J.K. Harry Potter and the Sorcerer's Stone. New York: Arthur A. Levine Books, 2015.

Rowling, J.K. Harry Potter and the Chamber of Secrets. Scholastic Inc., 1999.

Rowling, J.K. Harry Potter and the Order of Phoenix. Bloomsbury, 2004.

Rowling, J.K. Harry Potter and the Goblet of Fire. Scholastic Inc., 2002.

Rowling, J.K. Harry Potter and the Prisoner of Azkaban. Scholastic Inc., 2013.

Rowling, J.K. Harry Potter and the Deathly Hallows. New York: Arthur A. Levine Books, 2007.

Hi. I'm Sunny. For me, everything is an adventure. I am ready to try anything, take chances, see what happens - and help you try, too! I like to think I'm confident, caring and have an open mind. I will cheer for your success and encourage everyone! I'm ready to be a really good friend!

I've got a problem. Well, I've always got a problem. And I don't like it. It makes me cranky, and grumpy, impatient and the truth is, I got a bad attitude. There. I said it. I admit it. And the reason I feel this way? Math! I don't get it and it bums me out. Grrrr!

Not trying to brag, but I am the smartest Brainer that ever lived - and I'm a brilliant shade of blue. That's why they call me Smarty. I love to solve problems and I'm always happy to explain how things work - to help any Brainer out there! To me, work is fun, and math is a blast!

I scare easily. Like, even just a little ...Boo! Oh wow, I've scared myself! Anyway, they call me Pickles because I turn a little green when I get panicky. Especially with new stuff. Eek! And big complicated problems. Really any problem. Eek! There, I did it again.

Hi! Name's Pepper. I have what you call a positive outlook. I just think being alive is exciting! And you know something? By being friendly, kind and maybe even wise, you can have a pretty awesome day every day on this amazing planet.

A famous movie star once said, "I want to be alone." Well, I do too! I'm best when I'm dreaming, thinking, and in my own world. And so, I resist! Yes, I resist anything new, and only do things my way or quit. The rest of the Brainers have math, but I'd rather have a headache and complain. Or pout.

CHAPTER 1

Fun with

Multiplying and Dividing within 100 without Regrouping

Multiplying by 2-9s using arrays

Learning Multiplication Facts

Dividing by 2-9s

Learning the Order of Operations

Word Problems

Factors and Multiples

> You might belong in Gryffindor,
> Where dwell the brave at heart,
> Their daring, nerve and chivalry
> Set Gryffindors apart;
>
> (see *Harry Potter and the Sorcerer's Stone* page 97).

While you are at Hogwarts, your triumphs will earn your house points, while any rule-breaking will lose house points. At the end of the year, the house with the most points is awarded the House Cup, a great honor (see *Harry Potter and the Sorcerer's Stone* page 94).

One solved task gives you one "Triumph point", one mistake gives you one "Rule-breaking point". Welcome to your new year at Hogwarts!

Triumph points - …   Rule-breaking points - …

Everyone starts at the beginning at Hogwarts, ... I know it's hard. ... But yeh'll have a great time at Hogwarts - (see **Harry Potter and the Sorcerer's Stone** page 73).

What are these? (see **Harry Potter and the Sorcerer's Stone** page 85).

1	2	3	4	5	6	7	8	9	10
2	4	6	8	10	12	14	16	18	20

Finally! My favorite Multiplication-and-Division tables!

$$2 \times 7 = 7 + 7 = 14 \quad \text{or} \quad 18 \div 2 = 9.$$

Fill in the first row from 1 to 10.

Fill in the second row from 2 to 20 for 2's. Hint: multiply the base number 2 by each number in the first row.

1. Fill in the second rows from 3 to 30 for 3's, from 4 to 40 for 4's, and from 5 to 50 for 5's.

1	2	3	4	5	6	7	8	9	10
3	6	9	12	15	18	21	24	27	30

1	2	3	4	5	6	7	8	9	10
4	8	12	16	20	24	28	32	36	40

1	2	3	4	5	6	7	8	9	10
5	10	15	20	25	30	35	40	45	50

Triumph points - ... Rule-breaking points - ...

1. <u>Fill in</u> the Multiplication-and-Division tables.

...Imagine if they put me in Slytherin (see *Harry Potter and the Sorcerer's Stone* page 88).

1	2	3	4	5	6	7	8	9	10
6	12	18	24	30	36	42	48	54	60

1	2	3	4	5	6	7	8	9	10
7	14	21	28	35	42	49	56	63	70

1	2	3	4	5	6	7	8	9	10
8	16	24	32	40	48	56	64	72	80

1	2	3	4	5	6	7	8	9	10
9	18	27	36	45	54	63	72	81	90

2. <u>Help</u> me get the Triwizard Cup. Wow! Slytherin?!

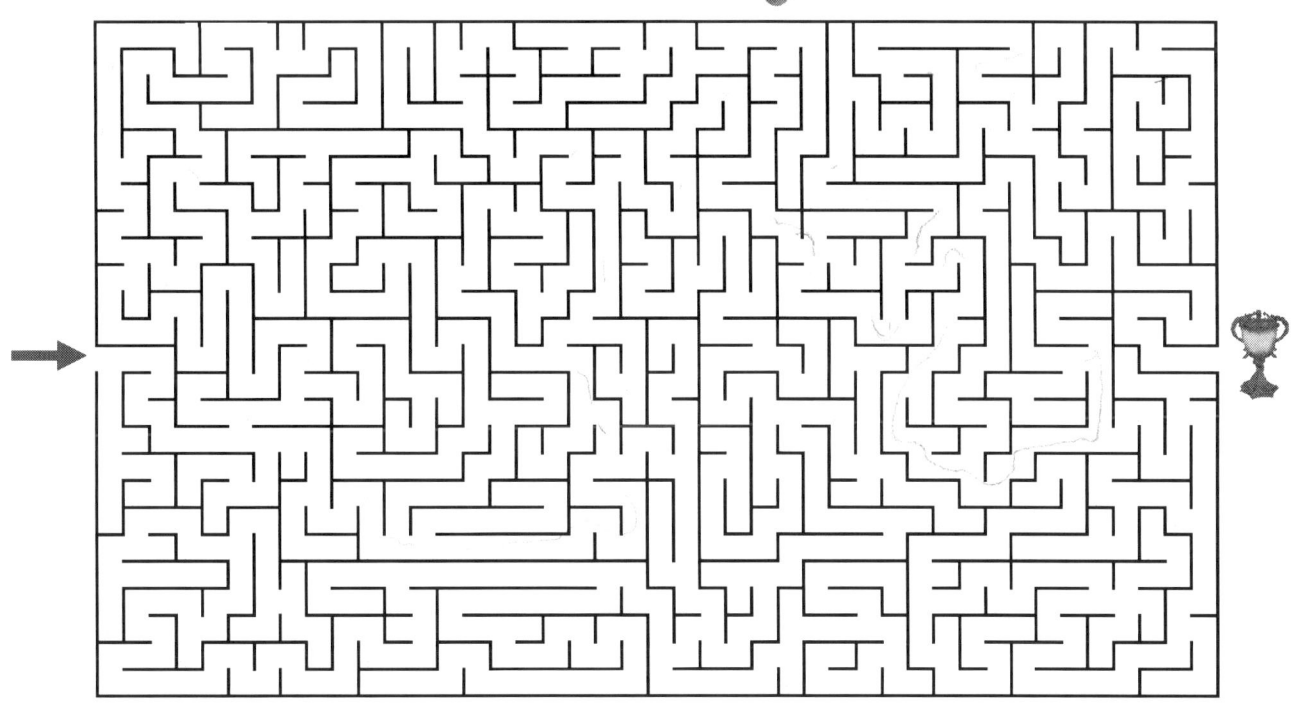

Triumph points - ...   Rule-breaking points - ...

How to multiply 4 × 8?

Step 1: find the product of two 8's: 2 × 8 = 16

Step 2: double the product: 16 + 16 = 32

4 × 8 = (2 × 8) + (2 × 8) = 32

1. Multiplysss.

4 × 7 = __28__   4 × 9 = __36__

4 × 4 = __16__   4 × 3 = __12__

4 × 5 = __20__   4 × 6 = __24__

How to multiply 6 × 8?

Step 1: find the product of two 8's: 2 × 8 = 16

Step 2: triple the product: 16 + 16 + 16 = 48

6 × 8 = (2 × 8) + (2 × 8) + (2 × 8) = 48

2. Multiply.

6 × 7 = __42__   6 × 9 = __54__

6 × 4 = __24__   6 × 3 = __18__

6 × 5 = __30__   6 × 6 = __36__

Triumph points - …   Rule-breaking points - …

1. <u>Write in</u> the missing numbers on a potion factor tree. The number shows how many bottles you need, and the bottle shows how many grams of herbs you need to make a potion.

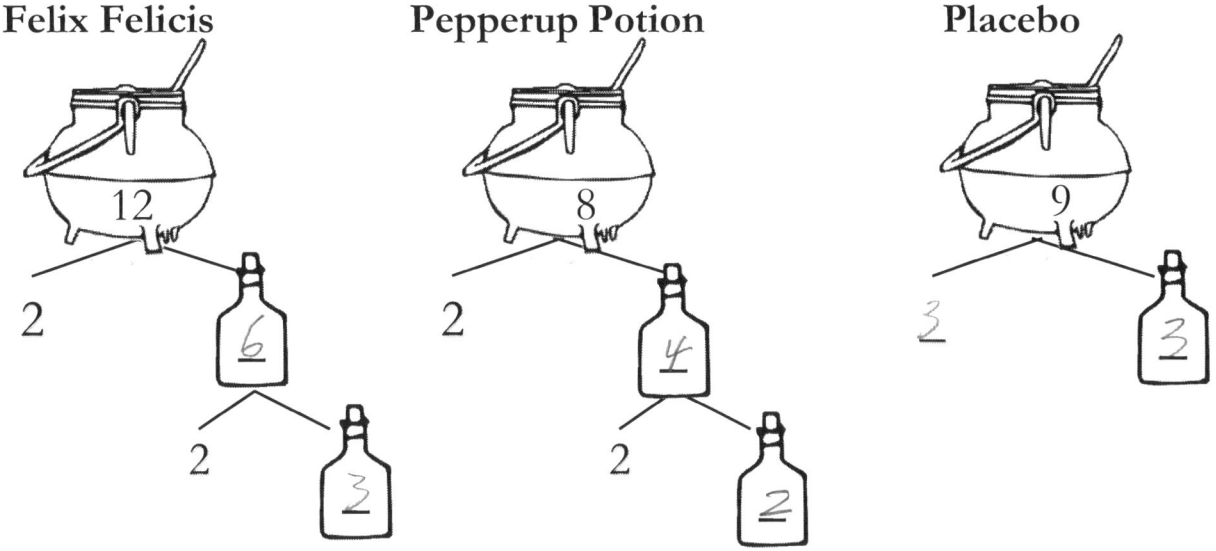

Factors are numbers that, when multiplied together (for example, 2 × 5) form a new number (for example, 10) called a product.

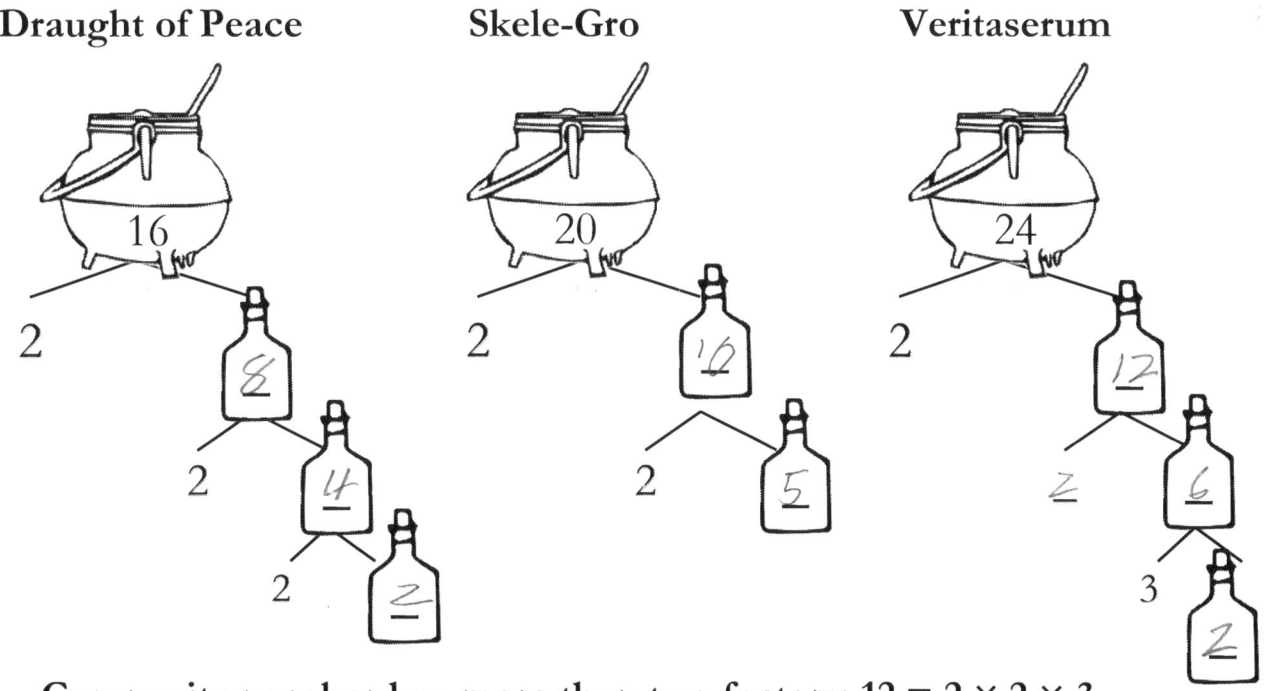

Composite number has more than two factors: 12 = 2 × 2 × 3.

How to multiply 7 × 8? Hint: when the multiplier (the first factor) is 1 less than the multiplicand (the second factor) you need to square the multiplicand (multiply the number by itself) and then, subtract the multiplicand out of the product:

Step 1: find the product of eight 8's: 8 × 8 = 64

Step 2: subtract one 8 out of the product: 64 - 8 = 56

7 × 8 = (8 × 8) - 8 = 56

1. <u>Multiply</u>. The multiplier (the first factor) is 1 less than the multiplicand (the second factor).

6 × 7 = __42__   8 × 9 = __72__

3 × 4 = __12__   2 × 3 = __6__

4 × 5 = __20__   9 × 10 = __90__

2. There are 35 pixies in all. I divided the pixies into groups of 5. How many small pixies are there?

__1__

How to multiply 6 × 5? Hint: when the multiplier (the first factor) is 1 more than the multiplicand (the second factor) you need to square the multiplicand (multiply the number by itself) and then, add one more multiplicand to the product:

Step 1: find the product of five 5's: 5 × 5 = 25

Step 2: add one 5 to the product: 25 + 5 = 30

6 × 5 = (5 × 5) + 5 = 30

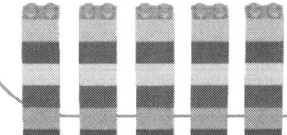

1. Multiply. The multiplier (the first factor) is 1 more than the multiplicand (the second factor).

5 × 4 = __20__   8 × 7 = __56__

3 × 2 = __6__   4 × 3 = __12__

9 × 8 = __72__   7 × 6 = __42__

2. There are 43 pixies in all. I divided the pixies into groups of 4. How many big pixies are there?

10 3/4

How to multiply 9 × 8?

Step 1: find the product of ten 8's: 10 × 8 = 80

Step 2: subtract one 8 out of the product: 80 - 8 = 72

9 × 8 = (10 × 8) - 8 = 72

1. Multiply.

9 × 7 = 63   9 × 9 = 81

9 × 4 = 36   9 × 3 = 27

9 × 5 = 45   9 × 6 = 54

How to multiply 5 × 8?

Hint: the best way to multiply by 5 is to multiply a number by 10 and divide the product by 2 since 5 = 10÷2

Step 1: find the product of ten 8's: 10 × 8 = 80

Step 2: divide the product by 2: 80 ÷ 2 = 40

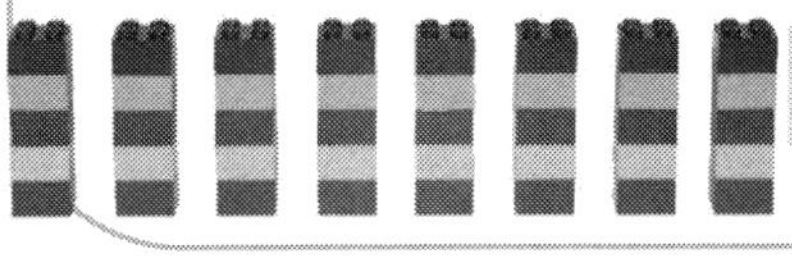

5 × 8 = (10 × 8) ÷ 2 = 40

2. Multiply.

5 × 7 = 35   5 × 9 = 45

5 × 4 = 20   5 × 3 = 15

5 × 5 = 25   5 × 6 = 30

Triumph points - …   Rule-breaking points - …

1. Find and circle or cross out the words to find out more about Harry Potter.

W	O	Y	U	D	D	T	F	I	A	Z	H	E	S	O	N	D
O	H	A	O	E	S	Z	G	K	L	A	O	L	I	O	N	E
A	I	I	R	F	F	P	W	R	R	U	F	D	G	R	R	I
O	T	M	T	I	L	C	I	R	O	S	Q	A	Z	U	V	R
N	A	R	Q	E	Y	A	Y	A	Y	Z	R	U	C	B	X	R
I	S	Y	T	Z	B	P	M	P	C	D	I	J	H	E	X	O
L	Q	W	A	R	O	U	F	O	G	I	I	B	O	U	J	W
L	G	T	P	T	L	L	M	N	C	R	R	B	G	S	T	M
D	R	V	T	U	X	C	I	B	N	A	E	S	D	H	J	A
B	D	E	T	C	L	Y	K	E	L	N	R	L	A	A	P	I
Q	R	C	A	U	F	S	F	E	F	E	X	D	J	G	Q	G
S	D	E	V	I	L	O	H	W	Y	O	B	M	U	R	Q	I
J	E	L	R	H	R	A	W	M	Z	R	P	E	L	I	C	O
H	P	R	Z	B	L	E	J	C	I	Z	X	M	E	D	D	U
V	E	A	L	B	U	S	D	U	M	B	L	E	D	O	R	E
T	W	U	R	G	W	J	Q	H	D	L	T	D	I	C	G	J
Q	P	A	S	X	A	G	X	C	K	K	T	U	Y	M	Q	W

HARRY POTTER

ALBUS DUMBLEDORE

RUBEUS HAGRID

I AM WORRIED

DRACO MALFOY

BOY WHO LIVED

WHITE BUMBLEBEE

I AM RED

TERRIFYING DRAGON

Triumph points - …

Rule-breaking points - …

> 1. Look at the pattern below and ansssswer the questionsss.

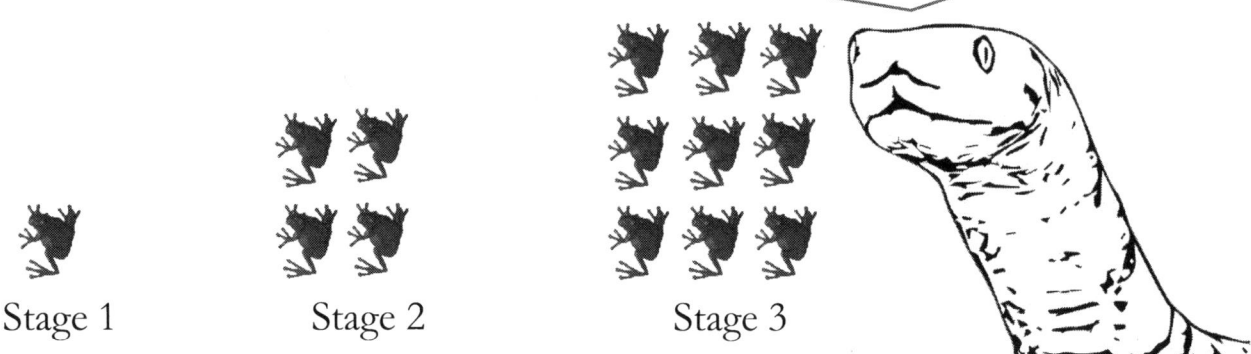

Stage 1   Stage 2   Stage 3

1. How many frogs are there on stage 4? _____.
2. How many frogs are there on stage 6? _____.
3. At what stage are there 64 frogs?   _____.
4. At what stage are there 25 frogs?   _____.

2. <u>Multiply</u>.

```
    4          5          7          6          8
×   2      ×   3      ×   4      ×   3      ×   7
   ───        ───        ───        ───        ───
    8         15         28         18         56

    2          5          8          7          3
×   6      ×   5      ×   9      ×   5      ×   8
   ───        ───        ───        ───        ───
   12         25         72         35         24

    6          6          9          9          6
×   4      ×   9      ×   3      ×   9      ×   7
   ───        ───        ───        ───        ───
   24         54         29         81         42
```

Triumph points - …   Rule-breaking points - …

1. Professor Lockhart showed two cages with pixies. The big cage had three times as many pixies as the small cage. If there were 27 pixies in the big cage, how many pixies were kept in the small cage?

Small cage	Big cage	Small cage
Unknown	27	? 9

Answer: _____

2. Write in the missing numbers on a potion factor tree. The number shows how many bottles you need, and the bottle shows how many grams of herbs you need to make a potion.

Ageing Potion

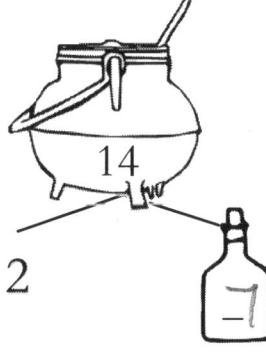

14
2   7

Hiccupping Potion

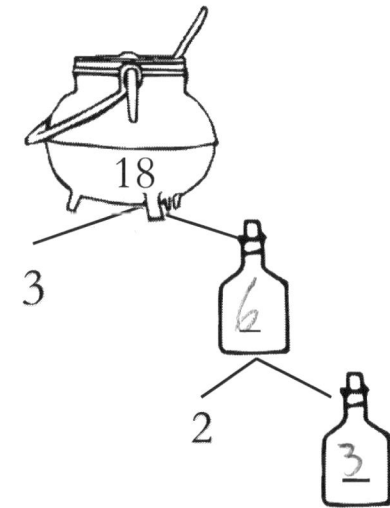

18
3   6
     2   3

Amortentia

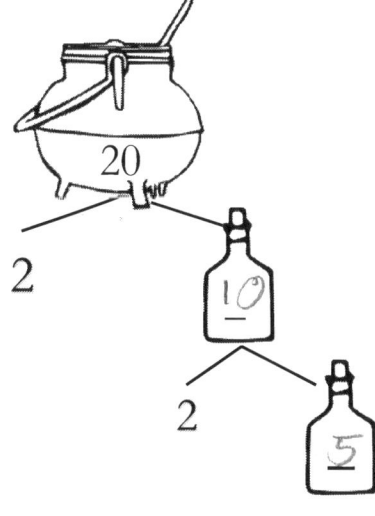

20
2   10
     2   5

Triumph points - …
Rule-breaking points - …

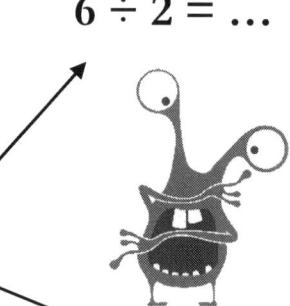

Brainers, I will do division. I like tricks, may I start with the tricks?

We divide 6 by 2:

$6 ÷ 2 = 3$.

$6 ÷ 2 = …$

6 divided by 2

Or I can rewrite the problem as: → $2\overline{)6}$

Step 1: divide 6 by 2: $6 ÷ 2 = 3$. The answer is 3. → $2\overline{)6}$

Step 2: write 3 in one's place above the division sign. → 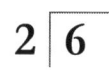 $2\overline{)6}$ (with 3 on top)

Step 3: multiply to check your answer: $3 × 2 = 6$. Write the product 6 under the dividend 6.

$$\begin{array}{r} 3 \\ 2\overline{)6} \\ -6 \end{array}$$

Step 4: subtract the product 6 from the dividend 6: $6 − 6 = 0$

$$\begin{array}{r} 3 \\ 2\overline{)6} \\ -6 \\ \hline 0 \end{array}$$

Triumph points - …

Rule-breaking points - …

1. Divide.

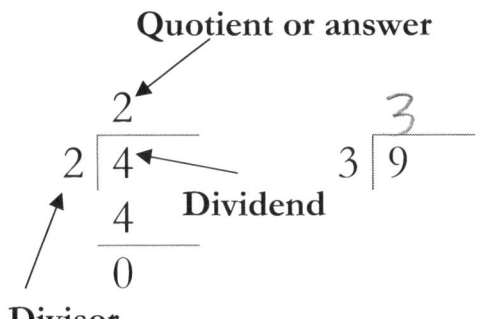

$3 \overline{) 9}$ quotient 3

$4 \overline{) 12}$ = 3

$5 \overline{) 10}$ = 2

$6 \overline{) 12}$ = 2

$8 \overline{) 16}$ = 2

$5 \overline{) 15}$ = 3

$7 \overline{) 14}$ = 2

$6 \overline{) 18}$ = 3

$5 \overline{) 25}$ = 5

$7 \overline{) 28}$ = 4

$8 \overline{) 24}$ = 3

$9 \overline{) 27}$ = 3

$8 \overline{) 32}$ = 4

$9 \overline{) 18}$ = 2

2. Ron got 8 points. Hermione got seven times as many points as Ron. How many points did they get altogether?

Ron	Hermione	Hermione	Altogether
8	7x	?56	?64

Answer: 64

Triumph points - … Rule-breaking points - …

1. <u>Find</u> the value.

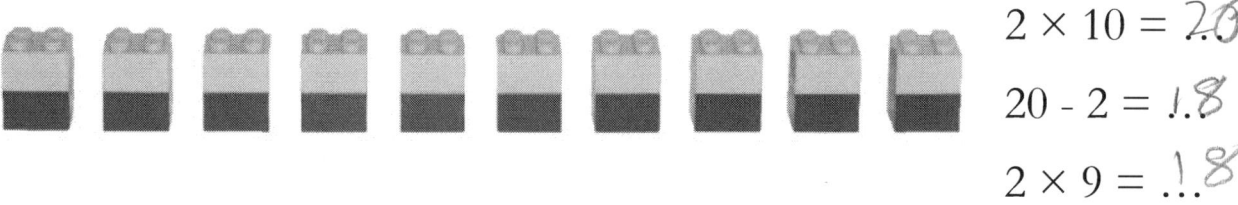

$2 \times 10 = 20$
$20 - 2 = 18$
$2 \times 9 = 18$

$5 \times 10 = 50$
$50 - 5 = 45$
$5 \times 9 = 45$

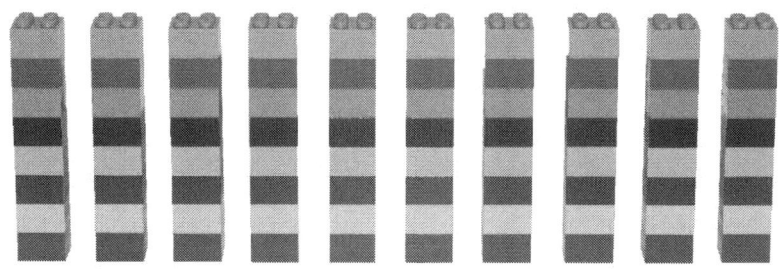

$8 \times 10 = 80$
$80 - 8 = 72$
$8 \times 9 = 72$

$3 \times 10 = 30$
$30 - 3 = 27$
$3 \times 9 = 27$

$9 \times 10 = 90$
$90 - 9 = 81$
$9 \times 9 = 81$

Triumph points - …   Rule-breaking points - …

1. <u>Multiply</u>. <u>Add</u> the products in each column. The Sorting Hat will choose the greatest sum. <u>Circle</u> the column with the greatest sum.

There's nothing hidden in your head The Sorting Hat can't see
(see *Harry Potter and the Sorcerer's Stone* page 97).

Not Slytherin, not Slytherin
(see *Harry Potter and the Sorcerer's Stone* page 100).

Slytherin	*Hufflepuff*	*Gryffindor*	*Ravenclaw*
7 × 4 = 28	8 × 5 = 40	6 × 3 = 18	5 × 5 = 25
3 × 4 = 12	7 × 2 = 14	9 × 3 = 27	2 × 8 = 16
3 × 7 = 21	4 × 6 = 24	5 × 3 = 15	8 × 7 = 56
9 × 5 = 45	8 × 4 = 32	9 × 9 = 81	8 × 9 = 72
9 × 6 = 54	8 × 6 = 48	7 × 7 = 49	4 × 2 = 8
5 × 4 = 20	6 × 5 = 30	2 × 5 = 10	8 × 5 = 40
2 × 6 = 12	7 × 5 = 35	3 × 6 = 18	6 × 6 = 36
2 × 9 = 18	4 × 9 = 36	8 × 3 = 24	7 × 3 = 21
4 × 4 = 16	6 × 4 = 24	5 × 5 = 25	5 × 3 = 15
7 × 6 = 42	9 × 7 = 63	2 × 2 = 4	8 × 8 = 64

Triumph points - …

Rule-breaking points - …

1. <u>Write in</u> the missing numbers on a potion factor tree. The number shows how many bottles you need, and the bottle shows how many grams of herbs you need to make a potion.

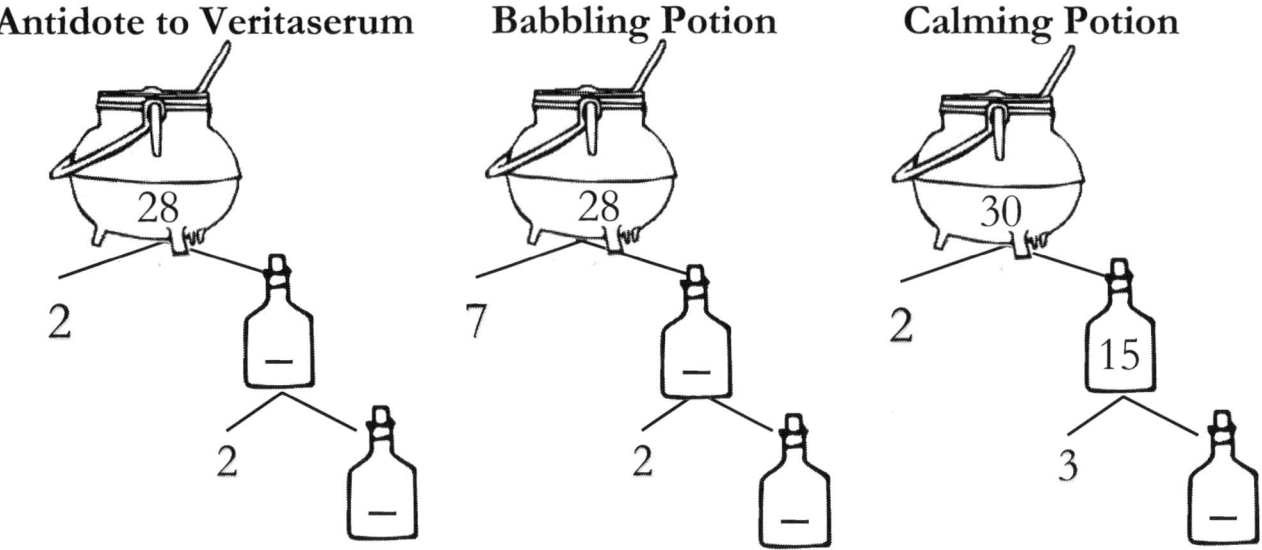

A prime number: is greater than 1; has factors only 1 and itself

2, 3, 5, 7, 11, 13, 17, 19, 23, 29, etc.

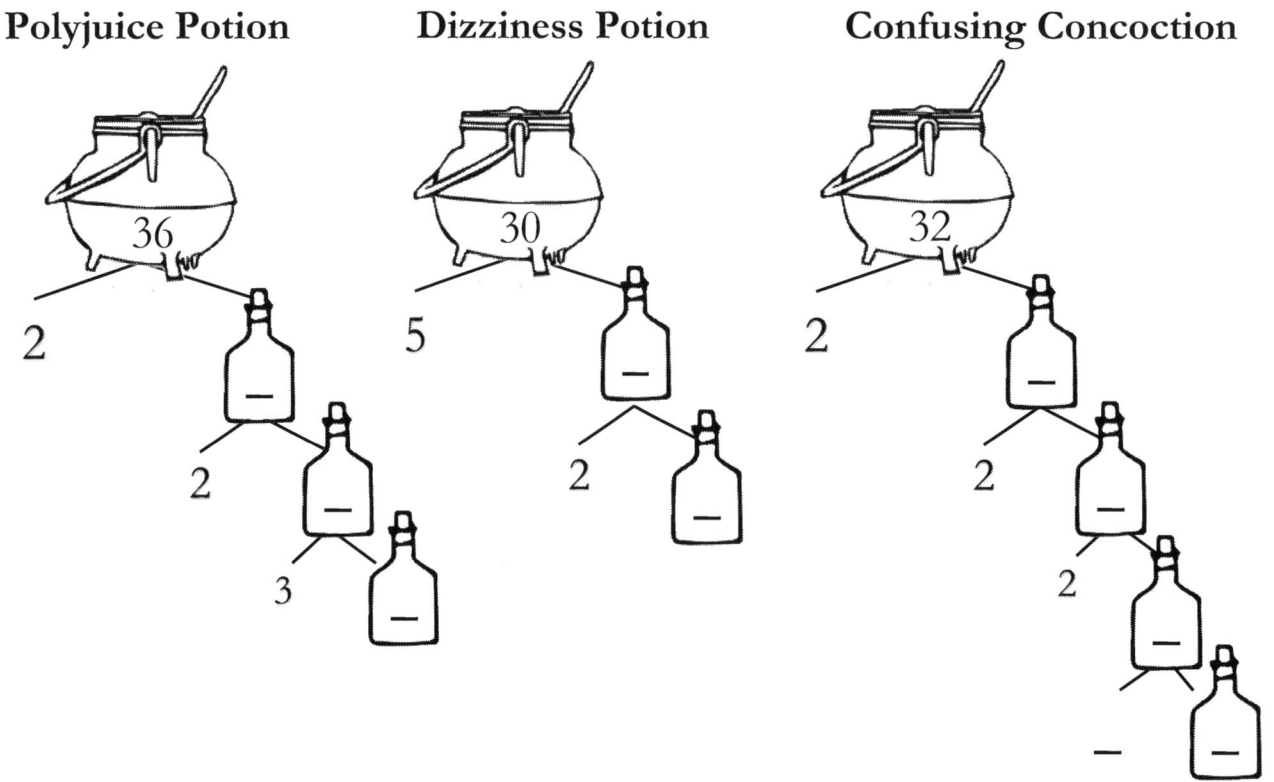

1. <u>Find</u> the value.

I bet I'm the worst in the class (see **Harry Potter and the Sorcerer's Stone** page 83).

$6 \times 10 = \ldots$

$60 - 6 = \ldots$

$6 \times 9 = \ldots$

$4 \times 10 = \ldots$

$40 - 4 = \ldots$

$4 \times 9 = \ldots$

$7 \times 10 = \ldots$

$70 - 7 = \ldots$

$7 \times 9 = \ldots$

2. <u>Evaluate</u> each expression. First, start with the parentheses (brackets), then, addition or subtraction from left to right! <u>Indicate</u> the order of operation.

$(7 \overset{1}{\times} 10) \overset{4}{+} (4 \overset{2}{\times} 9) \overset{5}{+} (8 \overset{3}{\times} 10) = \ldots \quad (7 \overset{1}{\times} 9) \overset{4}{+} (3 \overset{2}{\times} 10) \overset{5}{+} (8 \overset{3}{\times} 9) = \ldots$

$(9 \times 9) + (3 \times 9) + (6 \times 10) = \ldots \quad (5 \times 9) - (2 \times 10) + (6 \times 9) = \ldots$

$(5 \times 10) - (4 \times 10) + (2 \times 9) = \ldots \quad (9 \times 10) + (5 \times 9) - (6 \times 10) = \ldots$

Triumph points - …   Rule-breaking points - …

1. <u>Answer</u> the questions.

You won't be. There's loads of people who ... learn quick enough (see *Harry Potter and the Sorcerer's Stone* page 83).

I have a picture with black & white triangles. <u>How many white triangles</u> are there? Easy!

I use multiplication:

6 (down) × 3 (across) = 18 (white triangles).

Or I can add: 3 + 3 + 3 + 3 + 3 + 3 = 18.

<u>How many white squares</u> are there? <u>Write</u> one addition and one multiplication number sentence.

_____

_____

<u>How many white 4-pointed stars</u> are there?

_____

<u>How many black hexagons</u> are there?

_____

Triumph points - …   Rule-breaking points - …

Harry … could see the little round ball, wings fluttering, darting up ahead - he put on an extra spurt of speed -
(see *Harry Potter and the Sorcerer's Stone* page 155).

1. <u>Answer</u> the questions.

Find a pair of one-digit numbers when the sum is 14 and the product is 45: _____.

Find a pair of one-digit numbers when the sum is 16 and the product is 64: _____.

Triumph points - … Rule-breaking points - …

1. <u>Multiply</u> and <u>find</u> the value.

Brazil, here I come... Thanksss, amigo (see *Harry Potter and the Sorcerer's Stone* page 23).

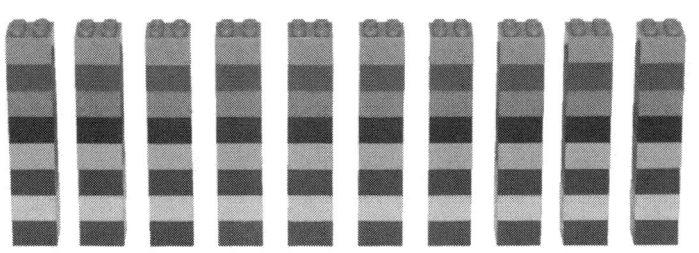

$8 \times 2 = \ldots$ $8 \times 4 = \ldots$ $8 \times 6 = \ldots$

$8 \times 8 = \ldots$ $8 \times 10 = \ldots$

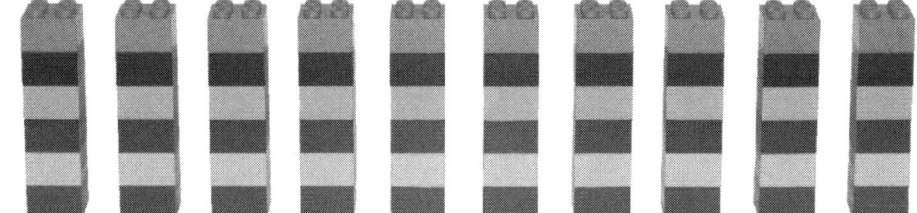

$6 \times 2 = \ldots$ $6 \times 4 = \ldots$ $6 \times 6 = \ldots$

$6 \times 8 = \ldots$ $6 \times 10 = \ldots$

2. <u>Evaluate</u> each expression. First, start with the parentheses (brackets), then, addition or subtraction from left to right! <u>Indicate</u> the order of operation.

$(8 \overset{1}{\times} 4) \overset{4}{+} (2 \overset{2}{\times} 2) \overset{5}{+} (5 \overset{3}{\times} 10) = \ldots$ $(2 \overset{1}{\times} 8) \overset{4}{+} (3 \overset{2}{\times} 6) \overset{5}{+} (8 \overset{3}{\times} 8) = \ldots$

$(3 \times 10) + (2 \times 6) + (8 \times 2) = \ldots$ $(5 \times 6) - (2 \times 10) + (3 \times 4) = \ldots$

$(5 \times 8) - (2 \times 4) + (3 \times 8) = \ldots$ $(8 \times 10) + (5 \times 2) - (3 \times 2) = \ldots$

$(5 \times 4) - (2 \times 2) + (8 \times 6) = \ldots$ $(8 \times 4) + (5 \times 4) - (5 \times 10) = \ldots$

Triumph points - ...   Rule-breaking points - ...

1. <u>Write in</u> the missing numbers on a potion factor tree. The number shows how many bottles you need, and the bottle shows how many grams of herbs you need to make a potion.

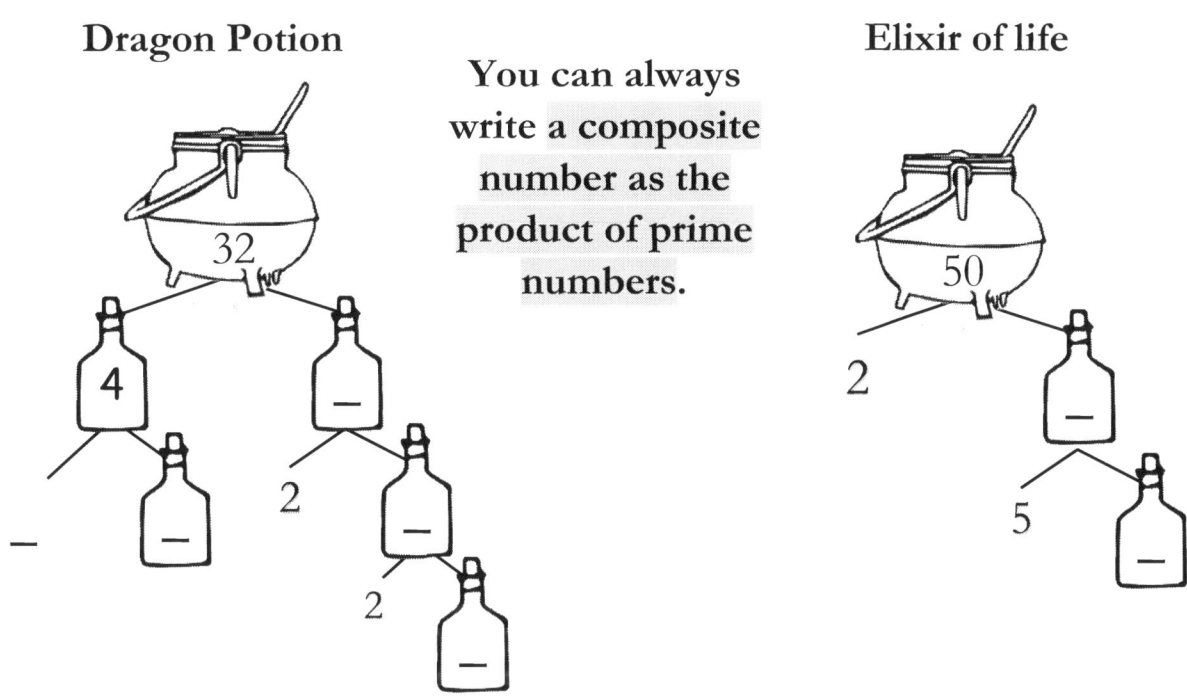

Dragon Potion

Elixir of life

You can always write **a composite number as the product of prime numbers.**

2. To brew a Draught of Living Death Potion you need to add powdered root of asphodel to an infusion of wormhood. The powdered root of asphodel weighs three times as much as an infusion of wormhood. If the powdered root of asphodel weighs 24 grams, <u>how much</u> do they weigh altogether?

Asphodel	Wormhood	Altogether
___	___	? ___

Answer: _____

Triumph points - … Rule-breaking points - …

1. <u>Answer</u> the questions. <u>Write</u> one multiplication and addition number sentence for the problem.

How many more green Christmas trees than red Christmas trees are there?

_____

<u>Find</u> the pattern horizontally or vertically. Hint: <u>find</u> the equal groups of objects; <u>multiply</u> the objects; and <u>add</u> the product. <u>Use</u> the easiest strategy.

How many stars are there?

$(2 \times 6) + (2 \times 8) + 10 = 12 + 16 + 10 = \ldots$

or use another pattern:

$(2 \times 4) + (5 \times 6) = 8 + 30 = \ldots$

I know some things. I can, you know, do maths and stuff

(see **Harry Potter and the Sorcerer's Stone** page 42).

Triumph points - …   Rule-breaking points - …

1. Find and circle or cross out the words to find out more about Harry Potter.

```
L H E L E N O F T R O Y I G R B G
I Y S S C O D S R O A Q Z T Q L E
D W O L F L I K E H U M A N W L O
S R Y G V G P V G O T B T C P T R
S O A K A K N N N A U N L U R F G
U R J Z U H H O A F I G U O P H E
A E M Z I E B T R L H O M E C D W
L M W R F W F X G K Y E J B O A E
E U L V H F D W E E D G K M Q D A
N S L X F G I O N L Q W U E F I S
E L D D I R O L O V R A M M O T L
M U A X X R B V I L C S L L F X E
G P K D H G D T M A B C Q R K U Y
N I Z V V R B R R G B E A K N Q O
I N D I O E F J E E O R R N P V G
K S J L T F E I H C S I M U J D B
T N U A G O L O V R A M E W P O I
```

TOM MARVOLO RIDDLE   LORD VOLDEMORT

MARVOLO GAUNT   PURE-BLOOD WIZARD

HERMIONE GRANGER   HELEN OF TROY

KING MENELAUS   BAILIFF

REMUS LUPIN   WOLFLIKE HUMAN

GEORGE WEASLEY   MISCHIEF

Triumph points - …   Rule-breaking points - …

1. <u>Look</u> at the pattern below and <u>answer</u> the questions.

Stage 1   Stage 2   Stage 3

1. <u>How many frogs and pixels</u> are there on stage 5? _____.

2. <u>How many frogs and pixels</u> are there on stage 10? _____.

3. <u>At what stage</u> are there 7 pixels? _____.

4. <u>At what stage</u> are there 49 frogs? _____.

2. <u>Multiply</u>.

```
    4          5          7          6          7
×   9      ×   6      ×   3      ×   6      ×   7
-----      -----      -----      -----      -----


    8          5          7          7          5
×   6      ×   9      ×   9      ×   6      ×   8
-----      -----      -----      -----      -----


    6          6          8          3          6
×   3      ×   9      ×   3      ×   9      ×   4
-----      -----      -----      -----      -----
```

Triumph points - …   Rule-breaking points - …

1. <u>Divide</u>.

$$\begin{array}{r}4\\2\overline{\smash{)}8}\\8\\\hline 0\end{array}\qquad 3\overline{\smash{)}21}\qquad 4\overline{\smash{)}36}\qquad 5\overline{\smash{)}40}\qquad 6\overline{\smash{)}30}$$

$$8\overline{\smash{)}56}\qquad 5\overline{\smash{)}30}\qquad 7\overline{\smash{)}42}\qquad 6\overline{\smash{)}36}\qquad 5\overline{\smash{)}45}$$

$$7\overline{\smash{)}63}\qquad 8\overline{\smash{)}72}\qquad 9\overline{\smash{)}54}\qquad 8\overline{\smash{)}64}\qquad 9\overline{\smash{)}81}$$

2. There are five times as many flying balls as both the tall goal posts and players on broomsticks in Quidditch. There are six tall goal posts and fourteen players on broomsticks. How many flying balls are in Quidditch?

Answer: _____.

Triumph points - …   Rule-breaking points - …

1. <u>Multiply</u> and <u>find</u> the value. <u>Indicate</u> order of operations.

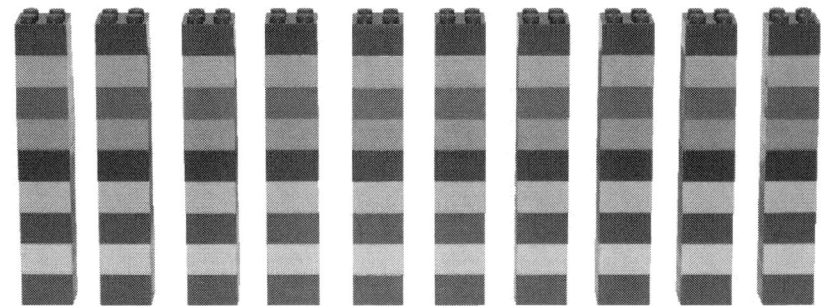

$9 \times 2 = \ldots$ $\qquad$ $9 \times 4 = \ldots$ $\qquad$ $9 \times 6 = \ldots$

$9 \times 8 = \ldots$ $\qquad$ $9 \times 10 = \ldots$

$7 \times 2 = \ldots$ $\qquad$ $7 \times 4 = \ldots$ $\qquad$ $7 \times 6 = \ldots$

$7 \times 8 = \ldots$ $\qquad$ $7 \times 10 = \ldots$

$(4 \overset{1}{\times} 4) \overset{4}{+} (9 \overset{2}{\times} 2) \overset{5}{+} (7 \overset{3}{\times} 10) = \ldots$ $\qquad$ $(6 \overset{1}{\times} 8) \overset{4}{+} (9 \overset{2}{\times} 4) \overset{5}{+} (4 \overset{3}{\times} 8) = \ldots$

$(6 \times 10) + (7 \times 6) + (4 \times 2) = \ldots$ $\qquad$ $(9 \times 6) - (4 \times 10) + (7 \times 4) = \ldots$

$(9 \times 8) - (4 \times 6) + (6 \times 2) = \ldots$ $\qquad$ $(9 \times 10) + (4 \times 2) - (7 \times 8) = \ldots$

$(6 \times 4) - (7 \times 2) + (9 \times 6) = \ldots$ $\qquad$ $(6 \times 6) + (7 \times 4) - (4 \times 10) = \ldots$

Triumph points - … $\qquad$ Rule-breaking points - …

> Dudley, meanwhile, was counting his presents. His face fell. "Thirty-six", (see Harry Potter and the Sorcerer's Stone page 17).

1. <u>Answer</u> the questions.

<u>Imagine</u> that Dudley got three kinds of different birthday presents (video games, racing cars, and remote-control aeroplanes). <u>Find</u> three one-digit numbers of 3 kinds of Dudley's birthday presents.

Note: any number has to be more than 1:

- the sum of these one-digit numbers is 10 and their product is 36:

_____

<u>Draw</u> the dots to show the pattern for the product and <u>circle</u> the bricks to show the number of Dudley's birthday presents.

Triumph points - …   Rule-breaking points - …

- the sum of these one-digit numbers is 11 and their product is 36:
_____

Draw the dots to show the pattern for the product and circle the bricks to show the number of Dudley's birthday presents.

- the sum of these one-digit numbers is 13 and their product is 36:
_____

Draw the dots to show the pattern for the product and circle the bricks to show the number of Dudley's birthday presents.

Dudley got 39 birthday presents. Place the parentheses to make these expressions correct.

5 + 6 × 4 − 5 = 39

47 − 24 ÷ 6 − 3 = 39

1. <u>Multiply.</u> <u>Add</u> the products in each column. The Triwizard Cup goes to the school with the greatest sum.

… become Hogwarts champion… was standing on the grounds… arms raised in triumph in front of the whole school… had just won the Triwizard Tournament (see *Harry Potter and the Goblet of Fire* page 192).

Hogwarts	*Beauxbatons*	*Durmstrang*
9 × 8 = …	8 × 3 = …	5 × 7 = …
9 × 5 = …	6 × 4 = …	7 × 3 = …
7 × 6 = …	4 × 8 = …	2 × 7 = …
7 × 9 = …	5 × 8 = …	8 × 7 = …
9 × 2 = …	6 × 6 = …	9 × 9 = …
8 × 2 = …	9 × 6 = …	8 × 8 = …
7 × 7 = …	4 × 5 = …	9 × 4 = …
6 × 5 = …	1 × 5 = …	8 × 5 = …
_____	_____	_____

Triumph points - … Rule-breaking points - …

1. <u>Answer</u> the questions. Waaaaaaaaaaa!

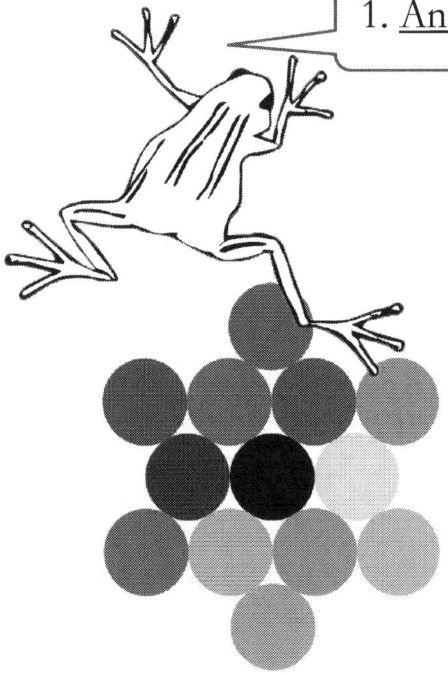

<u>How many circles</u> are there? <u>Write</u> one addition and multiplication number sentence for the problem.

____

____

____

____

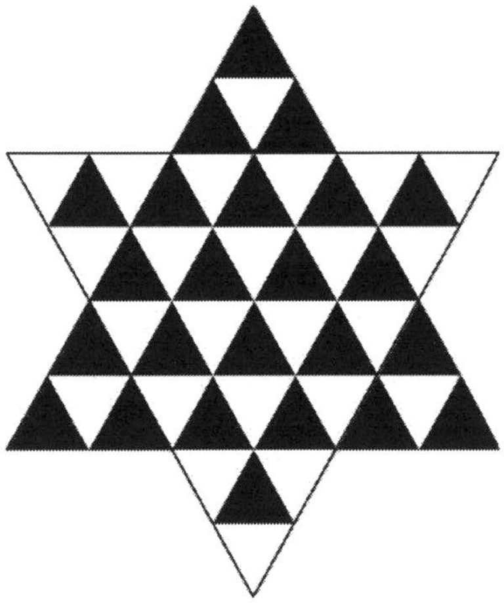

<u>How many more or less white triangles than black triangles</u> are there? <u>Use</u> the fastest strategy!

____

____

A

B

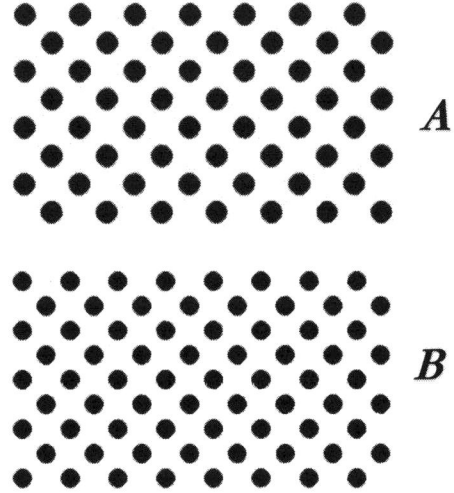

<u>Circle</u> the right answer.

1) A is greater than B

2) A is less than B

3) A is equal to B.

Triumph points - …

Rule-breaking points - …

1. <u>Multiply</u> and <u>find</u> the value. <u>Indicate</u> the order of operations.

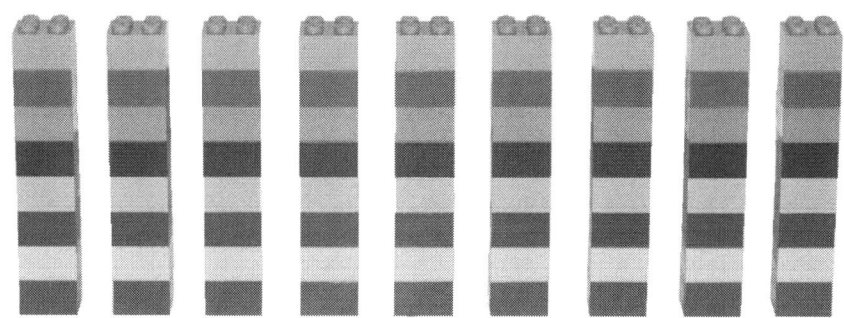

$8 \times 1 = \ldots$ $8 \times 3 = \ldots$ $8 \times 5 = \ldots$

$8 \times 7 = \ldots$ $8 \times 9 = \ldots$

... He hadn't learnt all the set books off by heart either (see *Harry Potter and the Sorcerer's stone* page 87).

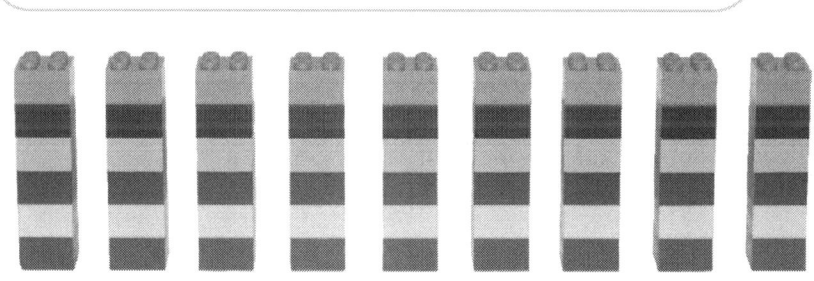

$6 \times 1 = \ldots$ $6 \times 3 = \ldots$ $6 \times 5 = \ldots$

$6 \times 7 = \ldots$ $6 \times 9 = \ldots$

$\overset{1}{(8 \times 5)} + \overset{4}{(2 \times 9)} + \overset{2}{(5 \times 5)} \overset{5}{} \overset{3}{} = \ldots$ $\overset{1}{(2 \times 5)} + \overset{4}{(3 \times 7)} + \overset{2}{(8 \times 9)} \overset{5}{} \overset{3}{} = \ldots$

$(3 \times 3) + (2 \times 3) + (8 \times 7) = \ldots$ $(5 \times 9) - (2 \times 7) + (3 \times 5) = \ldots$

$(8 \times 3) - (5 \times 3) + (3 \times 9) = \ldots$ $(8 \times 9) + (5 \times 7) - (2 \times 9) = \ldots$

Triumph points - ... Rule-breaking points - ...

1. <u>Answer</u> the questions. <u>Write</u> one multiplication and addition number sentence for each problem. <u>Circle</u> the groups of objects to show your fastest strategy.

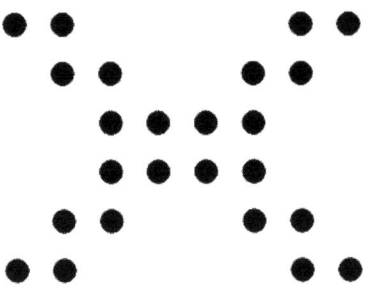

<u>How many dots</u> are there?

_____

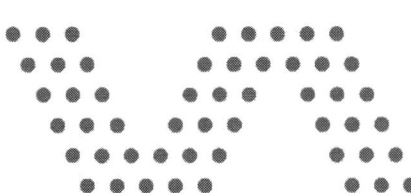

<u>How many dots</u> are there?

_____

*One more lesson like that and I might just do a **Weasley** (see **Harry Potter and the Order of the Phoenix** page 596).*

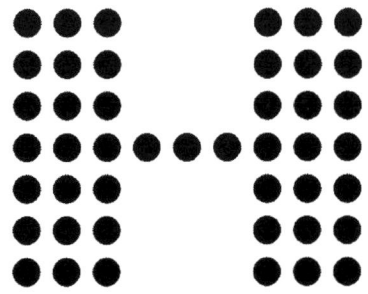

<u>How many dots</u> are there?

_____

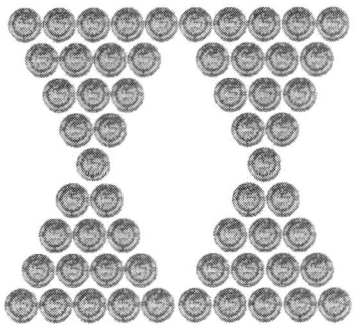

<u>How many bronze Knuts</u> are there?

_____

Triumph points - …

Rule-breaking points - …

1. <u>Find</u> and <u>circle</u> or <u>cross out</u> the words to find out more about Harry Potter.

```
S Z O W T A M S X L Q P P E G W S
W T B Y U L P J I A U B L Q R O G
B S E T Y D O L Y N I T B H Y L M
U F V R M I Y B X S R A N G F L F
I J J U C P U C E Y N Z N I F I Y
M P H A O E R Z M R M W Q A I W S
H E H T M P S G U J I A T S N G X
T V T X B E N F F Y H F R Y D N P
V E Q B Z I S Q O R B E A R O I D
R J T X N R Q P G R K O E A R P R
V D B A V F Q O O E E M B Z I M A
C V O C V G Q L E T B B C R Q O O
J M S Q O N R S P K T H M W S H B
P A R S E L T O N G U E D A Y W P
R A C G N I Y L F O A I R E H S U
S N I T C H J A E Q S W H T Q C C
S F Q B G H M K Y M N R E T Z C Y
```

GRYFFINDOR　　　　　　　FIREBOLT

LILY POTTER　　　　　　　JAMES POTTER

PARSELTONGUE　　　　　　SEEKER

CUPBOARD　　　　　　　　SNITCH

WHOMPING WILLOW　　　　FLYING CAR

MOANING MYRTLE　　　　　CHAMBER OF SECRETS

1. <u>Multiply</u>.

```
    2        5        8        6        6
×   9    ×   4    ×   3    ×   3    ×   7
———      ———      ———      ———      ———


    8        4        7        8        8
×   5    ×   9    ×   5    ×   6    ×   8
———      ———      ———      ———      ———


    6        9        7        5        6
×   4    ×   9    ×   3    ×   9    ×   5
———      ———      ———      ———      ———
```

2. George Weasley collected 72 Chocolate Frog cards with Albus Dumbledore. He collected 8 times as many cards as Fred Weasley. <u>How many cards</u> did Fred collect?

George	George	Fred
___	___	___
___		

Answer: _____

Triumph points - …

Rule-breaking points - …

1. <u>Divide</u>.

$2\overline{|16}$ = 8

$3\overline{|30}$ = 10
  -३↓
  00
  -00
  0

$4\overline{|40}$

$5\overline{|50}$

$6\overline{|60}$

$8\overline{|80}$

$5\overline{|20}$

$7\overline{|70}$

$6\overline{|12}$

$5\overline{|30}$

$7\overline{|35}$

$8\overline{|32}$

$9\overline{|45}$

$8\overline{|24}$

$9\overline{|72}$

2. Ron is thinking of a number. When this number is divided by 6, the answer is 9. <u>What</u> is the number?

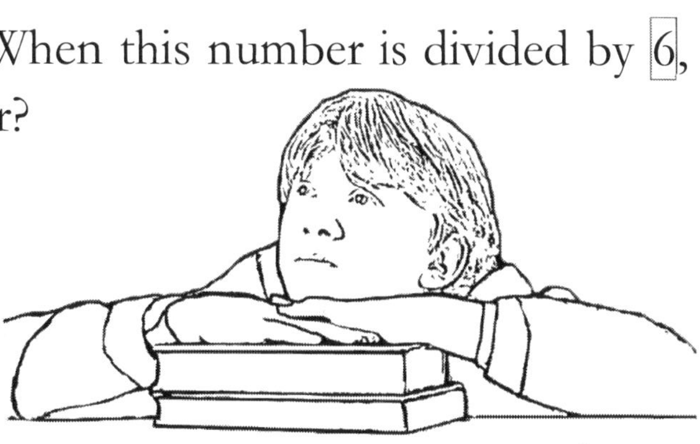

Answer:  .

1. Answer the questions.

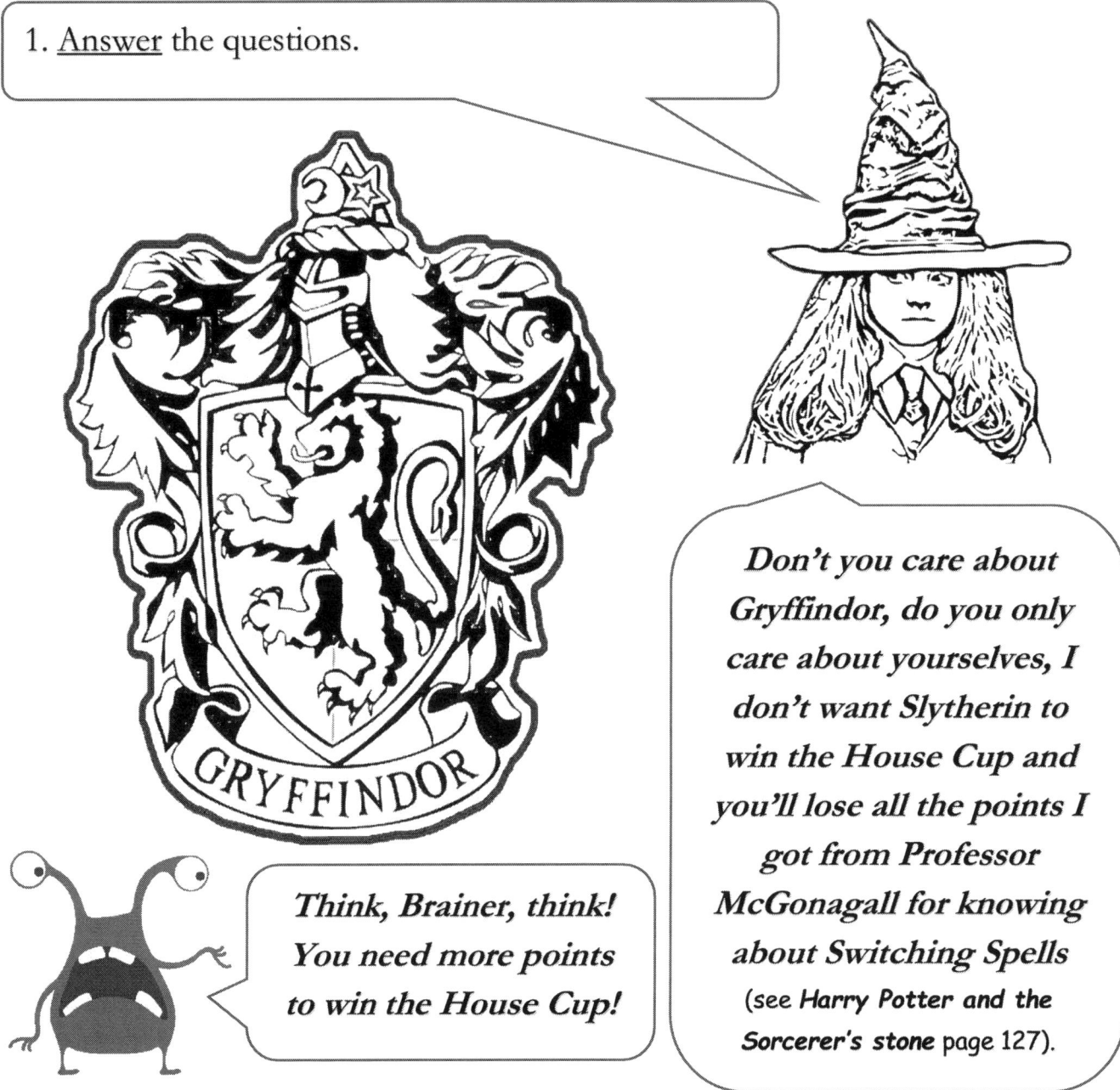

Think, Brainer, think! You need more points to win the House Cup!

Don't you care about Gryffindor, do you only care about yourselves, I don't want Slytherin to win the House Cup and you'll lose all the points I got from Professor McGonagall for knowing about Switching Spells (see *Harry Potter and the Sorcerer's stone* page 127).

Find a pair of two numbers: their difference is 12, and the product is 45:

_____.

Find a pair of two numbers: their difference is 7, and the product is 60:

_____.

Triumph points - …   Rule-breaking points - …

1. <u>Multiply</u> and <u>find</u> the value. <u>Indicate</u> the order of operations.

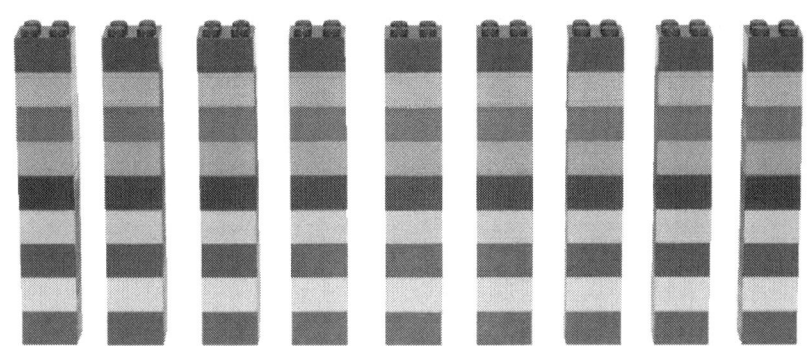

$9 \times 1 = \ldots$ $9 \times 3 = \ldots$ $9 \times 5 = \ldots$

$9 \times 7 = \ldots$ $9 \times 9 = \ldots$

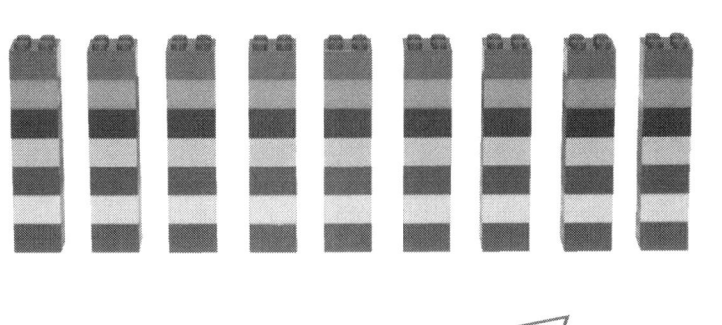

…you'll be just fine. Just be yerself. I know it's hard. (see **Harry Potter and the Sorcerer's stone** page 73).

$7 \times 1 = \ldots$ $7 \times 3 = \ldots$ $7 \times 5 = \ldots$

$7 \times 7 = \ldots$ $7 \times 9 = \ldots$

$(\overset{1}{4 \times 5}) + (\overset{4}{7 \times 9}) + (\overset{2}{6 \times 5}) \overset{5}{+} (\overset{3}{6 \times 5}) = \ldots$ $(7 \times 5) + (6 \times 7) + (4 \times 9) = \ldots$

$(7 \times 3) + (9 \times 3) + (4 \times 7) = \ldots$ $(6 \times 9) - (7 \times 7) + (9 \times 5) = \ldots$

$(6 \times 3) - (4 \times 3) + (9 \times 9) = \ldots$ $(9 \times 7) + (4 \times 3) - (6 \times 5) = \ldots$

Triumph points - …   Rule-breaking points - …

1. <u>Answer</u> the questions. <u>Write</u> one multiplication and addition number sentence for each problem. <u>Circle</u> the groups of objects to show your strategy.

So what are we going to do? (see Harry Potter and the Deathly Hallows page 47).

No magic! Look for patterns or multiplication arrays!

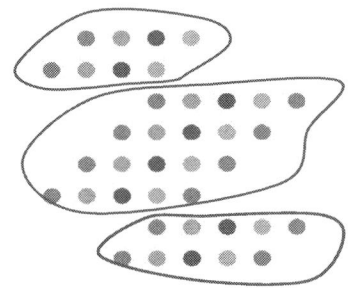

How many dots are there? <u>Use</u> the best strategy.

$(2 \times 4) + (4 \times 5) + (2 \times 5) = \ldots$

How many dots are there?

_____

_____

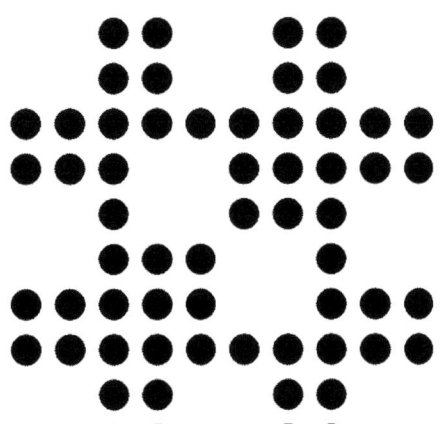

How many dots are there?

_____

_____

Triumph points - …   Rule-breaking points - …

1. <u>Multiply</u> and <u>find</u> the value. <u>Indicate</u> the order of operations. That's too easy!

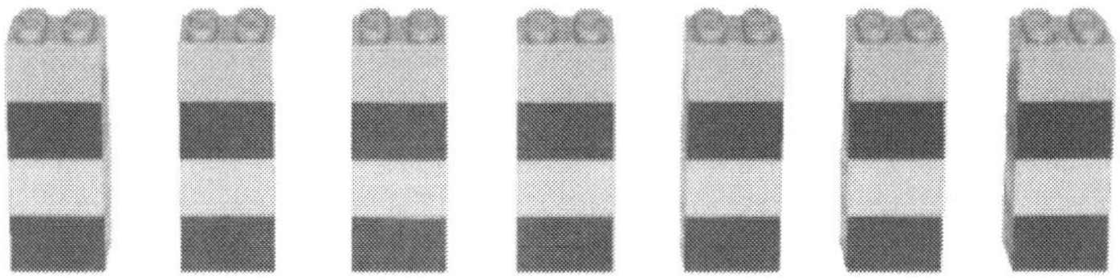

$4 \times 5 = \ldots$ $4 \times 2 = \ldots$

$4 \times 7 = \ldots$

1 4 2 5 3

$(4 \times 2) + (4 \times 7) + (4 \times 5) = \ldots$

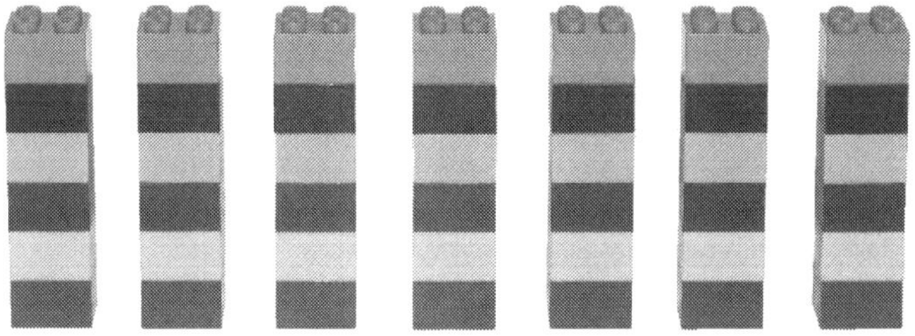

$6 \times 5 = \ldots$ $6 \times 2 = \ldots$

$6 \times 7 = \ldots$

$(6 \times 5) + (6 \times 7) + (6 \times 2) = \ldots$

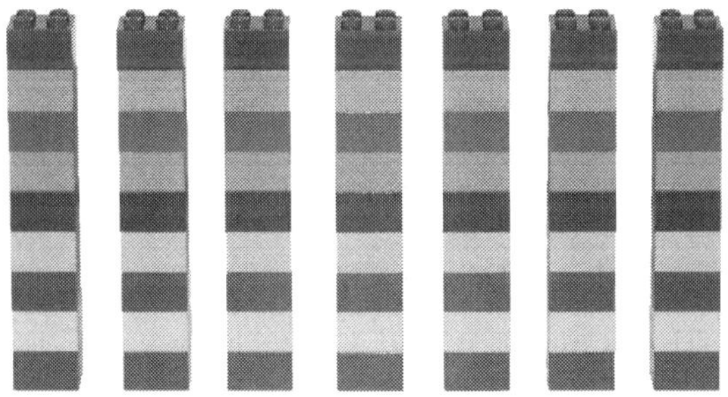

$9 \times 5 = \ldots$ $9 \times 2 = \ldots$

$9 \times 7 = \ldots$

$(9 \times 2) + (9 \times 7) + (9 \times 5) = \ldots$

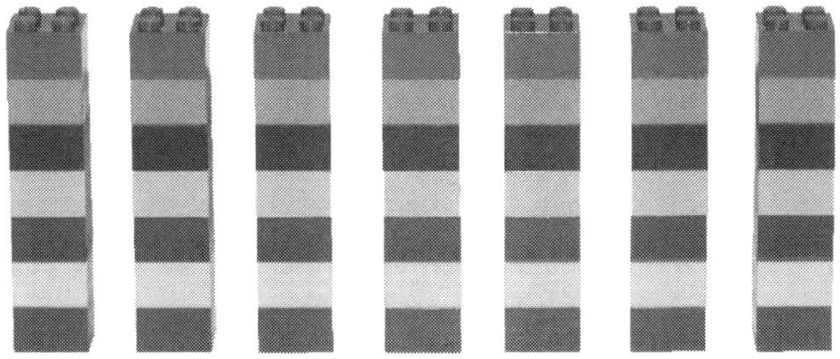

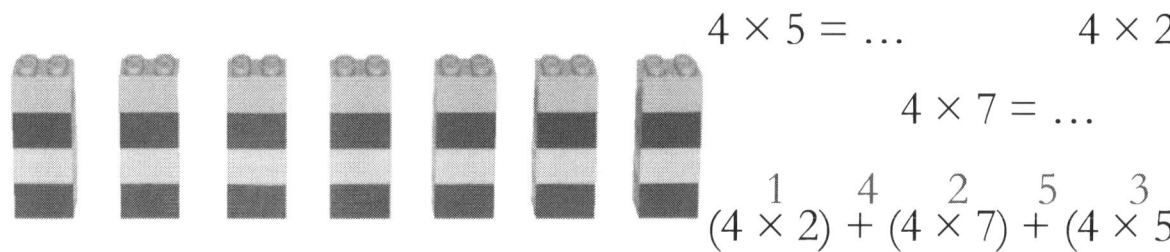

$7 \times 5 = \ldots$ $7 \times 2 = \ldots$

$7 \times 7 = \ldots$

$(7 \times 5) + (7 \times 7) + (7 \times 2) = \ldots$

$(4 \times 5) + (7 \times 2) + (6 \times 5) = \ldots$ $(7 \times 5) + (6 \times 7) + (4 \times 2) = \ldots$

$(7 \times 7) + (9 \times 2) + (4 \times 7) = \ldots$ $(9 \times 5) - (6 \times 2) + (9 \times 7) = \ldots$

Triumph points - … Rule-breaking points - …

1. <u>Write in</u> the missing numbers on a potion factor tree. The number shows how many bottles you need, and the bottle shows how many grams of herbs you need to make a potion.

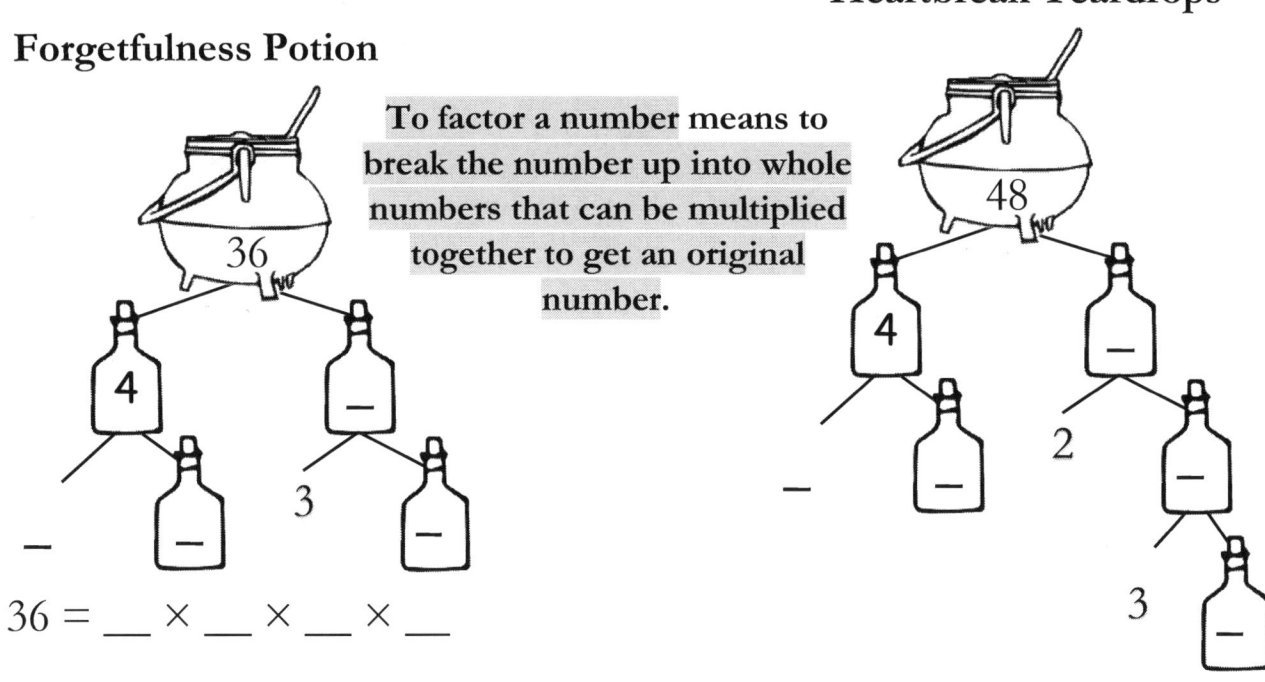

To factor a number means to break the number up into whole numbers that can be multiplied together to get an original number.

36 = __ × __ × __ × __

48 = __ × __ × __ × __ × __

Divisor is the number by which you're dividing

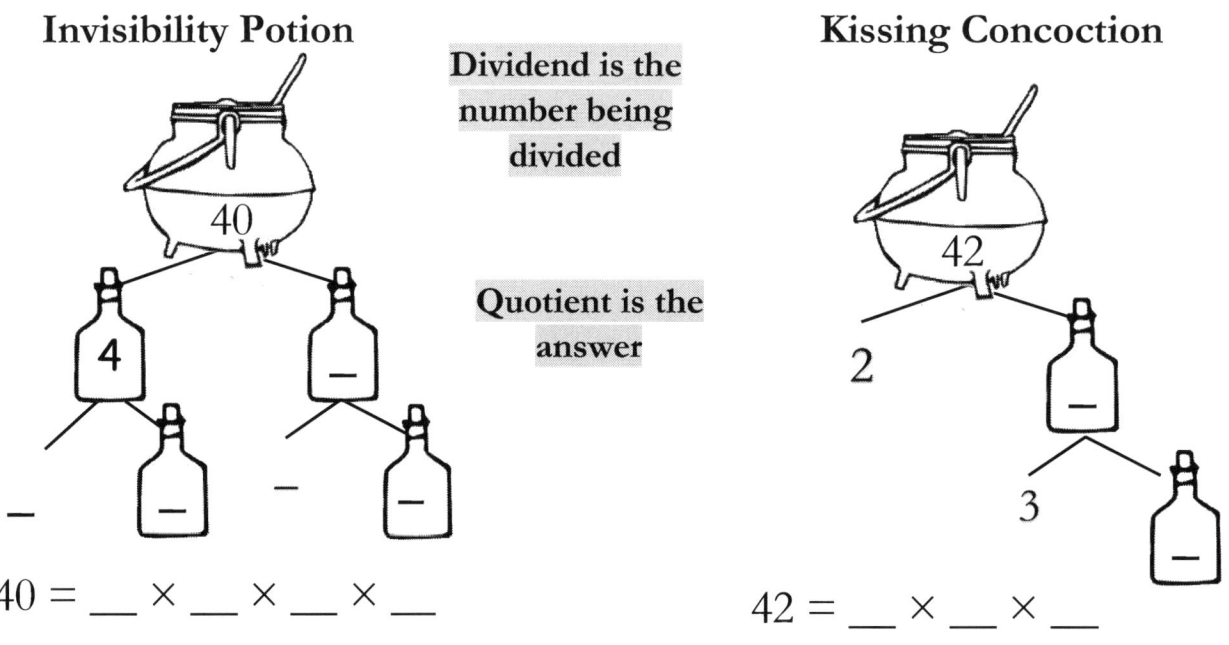

Dividend is the number being divided

Quotient is the answer

40 = __ × __ × __ × __

42 = __ × __ × __

Triumph points - …

Rule-breaking points - …

1. Circle the right answer.

A	3 × 4 × 7	A is equal to B
B	2 × 7 × 6	A is greater than B
		A is less than B

A	(8 × 7) - 19	A is greater than B
B	(9 × 6) - 28	A is equal to B
		A is less than B

A	(9 × 3) + 14	A is greater than B
B	(7 × 5) + 28	A is equal to B
		A is less than B

A	2 × 2 × 6 × 3	A is greater than B
B	4 × 2 × 3 × 2	A is less than B
		A is equal to B

Triumph points - …   Rule-breaking points - …

1. <u>Multiply</u> and <u>find</u> the value. <u>Indicate</u> the order of operations. Ca-ha-ha-ac!

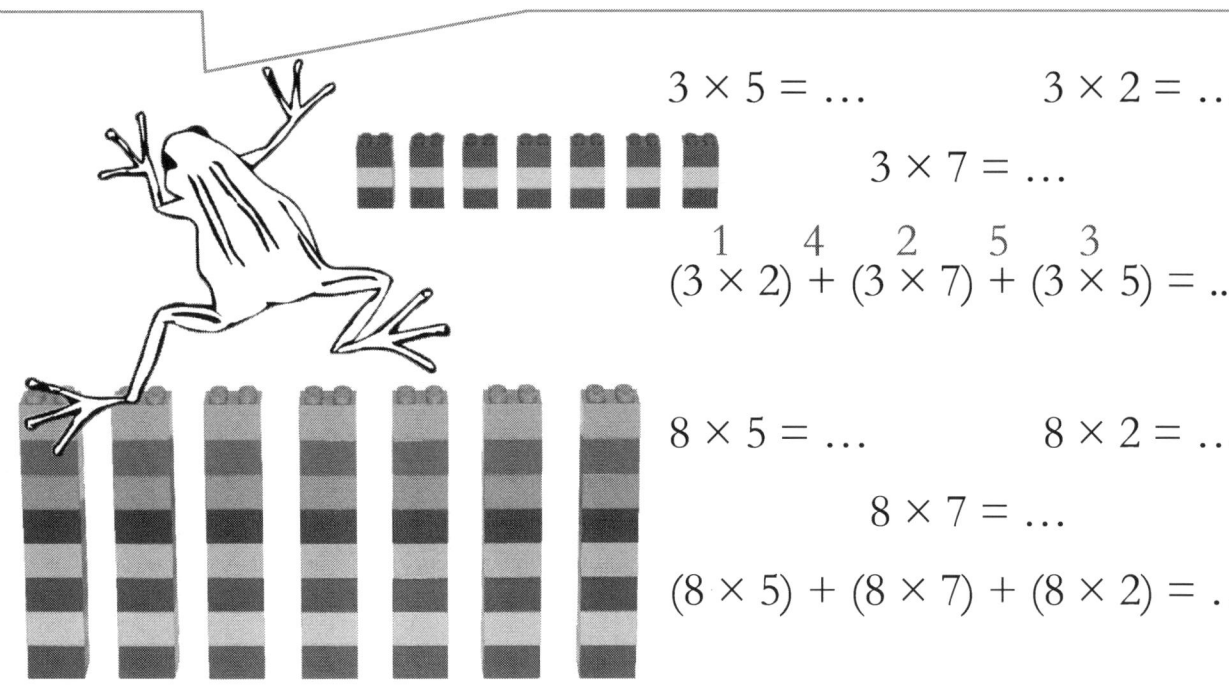

$3 \times 5 = \ldots \qquad 3 \times 2 = \ldots$

$3 \times 7 = \ldots$

$\overset{1}{(3 \times 2)} + \overset{4}{(3 \times 7)} + \overset{2}{(3 \times 5)} \overset{5}{=} \overset{3}{\ldots}$

$8 \times 5 = \ldots \qquad 8 \times 2 = \ldots$

$8 \times 7 = \ldots$

$(8 \times 5) + (8 \times 7) + (8 \times 2) = \ldots$

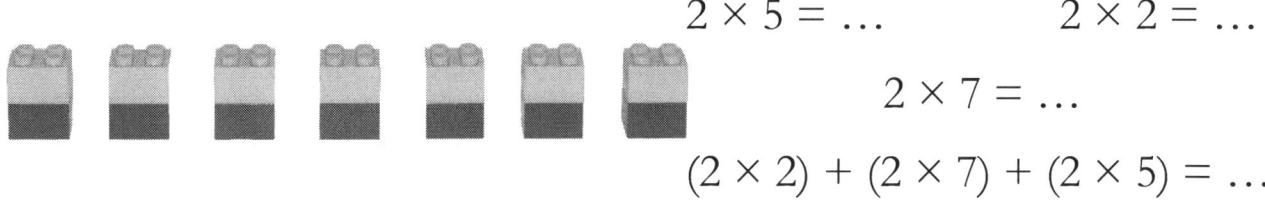

$2 \times 5 = \ldots \qquad 2 \times 2 = \ldots$

$2 \times 7 = \ldots$

$(2 \times 2) + (2 \times 7) + (2 \times 5) = \ldots$

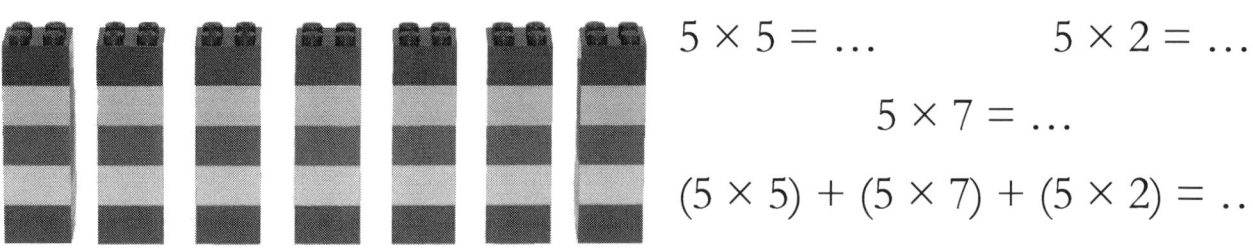

$5 \times 5 = \ldots \qquad 5 \times 2 = \ldots$

$5 \times 7 = \ldots$

$(5 \times 5) + (5 \times 7) + (5 \times 2) = \ldots$

$(3 \times 5) + (2 \times 2) + (8 \times 5) = \ldots \qquad (2 \times 5) + (5 \times 7) + (3 \times 2) = \ldots$

$(5 \times 2) + (8 \times 2) + (3 \times 7) = \ldots \qquad (8 \times 7) - (5 \times 5) + (2 \times 7) = \ldots$

Triumph points - … \qquad\qquad Rule-breaking points - …

1. <u>Answer</u> the questions. <u>Write</u> one multiplication and addition, and one division number sentence for each problem. <u>Fill in</u> the missing numbers.

I divided the dots into ⑤ groups, <u>how many dots</u> do I have in each group?

Step 1: dots in all: (4×6)+(3×7)=24+21=…

Step 2: dots per group: … ÷ 5 = …

I divided the stars into ⑤ groups, <u>how many stars</u> do I have in each group?

_____

_____

I divided the stars into ⑥ groups, <u>how many stars</u> do I have in each group?

_____

_____

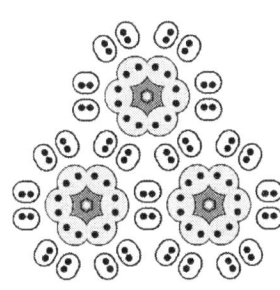

I divided the black dots into ③ groups, <u>how many dots</u> do I have in each group?

_____

_____

Triumph points - …

Rule-breaking points - …

1. <u>Find</u> and <u>circle</u> or <u>cross out</u> the words to find out more about Professor Dumbledore.

```
X F B K E D E J F K H N I A G O R O
I T M W E U P A L A J Y D Y V S I Y
N I H I X U K K U Q W L M C C E Z U
E X J S E J Z V Z I H K W I E B N N
O L B D N X H S G X V V E B E I M Q
H K A O L C Y T I L I B I S I V N I
P B D M P I H S D N E I R F Z X I G
E N O M E L T E B R E H S D B C R C
H S Y U V N Q V O V N B U V P O B U
T C H O C O L A T E F R O G C A R D
F K N I T T I N G P A T T E R N S M
O V Z S K L X Z V V I M V K A L U R
R R X W O O L Y S O C K S V Z S V Q
E T B U R N I N G D A Y M Y I T D J
D J M D V W O N Z K D Q Q C Z F D C
R E T S A M D A E H S T R A W G O H
O V G O O S Y N M Q J V F T E E C L
H Q P S S B Q W N W N K J C U D X P
```

HOGWARTS HEADMASTER

SHERBET LEMON

ORDER OF THE PHOENIX

BURNING DAY

CHOCOLATE FROG CARD

MUSIC

FAWKES

FRIENDSHIP

WOOLY SOCKS

INVISIBILITY CLOAK

KNITTING PATTERNS

WISDOM

1. Multiply.

```
    2        5        7        6        9
×   5    ×   8    ×   3    ×   6    ×   7
___      ___      ___      ___      ___

    8        6        7        4        7
×   4    ×   9    ×   4    ×   6    ×   8
___      ___      ___      ___      ___

    4        9        7        8        6
×   4    ×   6    ×   5    ×   9    ×   3
___      ___      ___      ___      ___
```

2. There are 6 trees along the road in front of Hagrid's house. The trees are 2 ft away from each other. How far (in inches) is the third tree from the first tree?

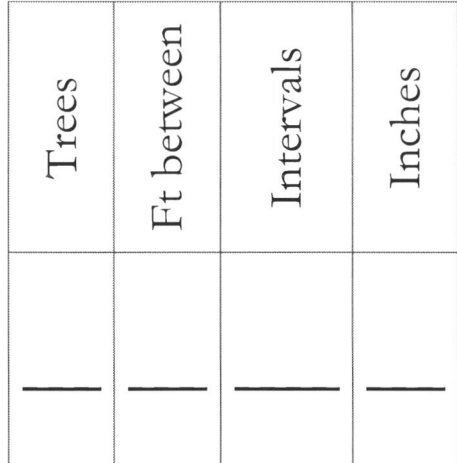

Trees	Ft between	Intervals	Inches
___	___	___	___

Answer: _____

1. Mr. Dursley saw many people whispering excitedly and dressed in cloaks that day. The number of people he met in the morning was four times the number of people he met during the lunch. There were 40 people in cloaks altogether. How many more or less people in cloaks in the morning than in the afternoon were there?

Morning	Lunch	In all	Morning > or < Lunch
_____	? ___	___	___

Answer: _____

2. Write in the missing numbers on a potion factor tree. The number shows how many bottles you need, and the bottle shows how many grams of herbs you need to make a potion.

Laugh Inducing Potion

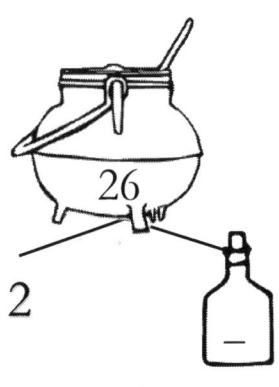

26 = __ × __

Sleeping Potion

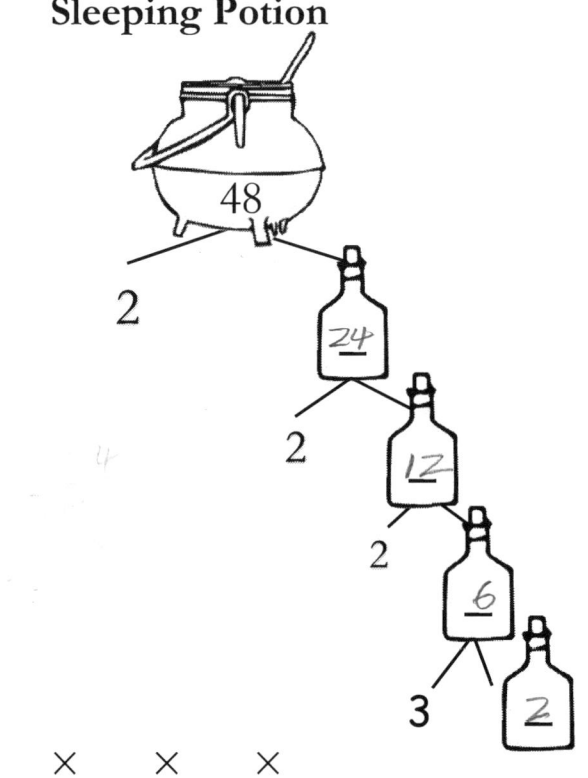

48 = __ × __ × __ × __ × __

Triumph points - …

Rule-breaking points - …

1. <u>Divide</u>.

$2\overline{|18}$ $3\overline{|24}$ $4\overline{|36}$ $5\overline{|30}$ $6\overline{|54}$

$8\overline{|40}$ $5\overline{|45}$ $7\overline{|21}$ $6\overline{|42}$ $5\overline{|50}$ 10

$7\overline{|49}$ $8\overline{|56}$ $9\overline{|63}$ $8\overline{|16}$ $9\overline{|27}$

2. Mr. Dursley counted 21 owls in the morning. There were two times as many flying owls as sitting owls on the trees. <u>How many owls</u> were flying?

42

Triumph points - …   Rule-breaking points - …

1. How much will Dobby earn a year?

_____

1. <u>Multiply</u> and <u>find</u> the value. <u>Indicate</u> the order of operations.

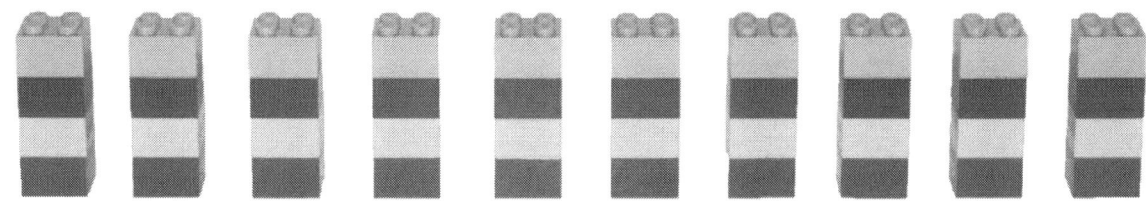

$4 \times 2 = \ldots$ $\quad 4 \times 6 = \ldots \quad$ $4 \times 8 = \ldots \quad 4 \times 10 = \ldots$

$\quad\quad\quad\quad\quad\; 1 \quad\; 4 \quad\; 2 \quad\; 5 \quad\; 3$
$(4 \times 2) + (4 \times 10) + (4 \times 8) = \ldots$

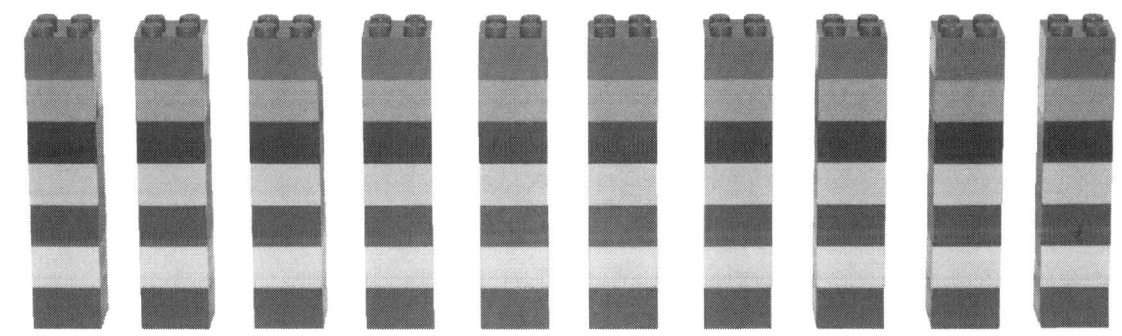

$7 \times 2 = \ldots \quad 7 \times 4 = \ldots \quad 7 \times 8 = \ldots \quad 7 \times 10 = \ldots$

$(7 \times 10) + (7 \times 8) + (7 \times 2) = \ldots$

$9 \times 2 = \ldots \quad 9 \times 4 = \ldots \quad 9 \times 8 = \ldots \quad 9 \times 10 = \ldots$

$(9 \times 2) + (9 \times 8) + (9 \times 10) = \ldots$

Triumph points - … Rule-breaking points - …

1. <u>Multiply</u>. <u>Use</u> five multiplication number sentences for each problem (decompose the factors). The first one is done for you.

8 × 8 = 4 × 2 × 4 × 2 = 1 × 8 × 8 × 1 = 2 × 2 × 2 × 2 × 2 × 2 =
= 16 × 4 = 8 × 4 × 2 = 64

9 × 8 = _____

_____

10 × 6 = _____

_____

8 × 6 = _____

_____

2. <u>Help</u> me get the Triwizard Cup. Don't leave me alone! Help needed!

Triumph points - … Rule-breaking points - …

1. <u>Multiply</u> and <u>find</u> the value. <u>Indicate</u> the order of operations.

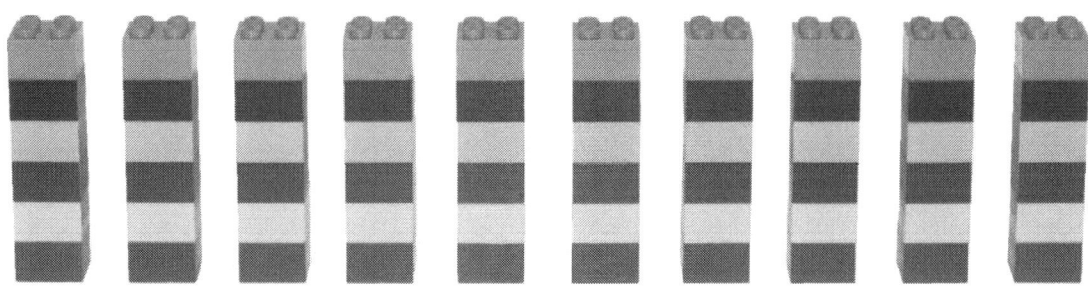

6 × 2 = …   6 × 4 = …   6 × 8 = …   6 × 10 = …

(6 × 8) + (6 × 10) + (6 × 2) = …

(4 × 10) + (9 × 2) + (7 × 8) = …   (9 × 10) + (6 × 8) + (4 × 2) = …

(6 × 10) + (7 × 10) + (4 × 8) = … (7 × 2) - (6 × 2) + (9 × 8) = …

3 × 2 = …   3 × 4 = …   3 × 8 = …   3 × 10 = …

(3 × 2) + (3 × 10) + (3 × 8) = …

2 × 2 = …   2 × 4 = …   2 × 8 = …   2 × 10 = …

(2 × 2) + (2 × 8) + (2 × 10) = …

Triumph points - …   Rule-breaking points - …

1. <u>Answer</u> the questions. <u>Write</u> one multiplication and addition, and one division number sentence for each problem.

I need to divide the dots into groups of 9, <u>how many groups</u> will I have?

_____

_____

I need to divide the dots into groups of 3, <u>how many groups</u> will I have?

Step 1: dots in all: (5×9)+(4×9)=45+36=…

Step 2: groups of 3 dots: … ÷ 3 = …

It does not do to dwell on dreams and forget to live (see *Harry Potter and the Sorcerer's Stone* page 174).

So, guys, let's do math!

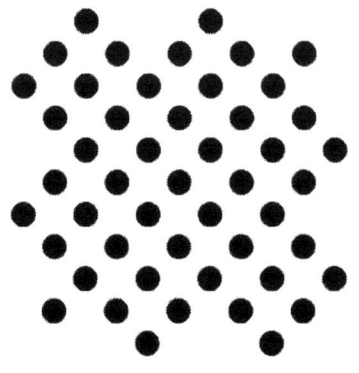

I need to divide the dots into groups of 7, <u>how many groups</u> will I have?

_____

_____

I need to divide the circles into groups of 8, <u>how many groups</u> will I have?

_____

_____

Triumph points - …

Rule-breaking points - …

1. <u>Write in</u> the missing numbers on a potion factor tree. The number shows how many bottles you need, and the bottle shows how many grams of herbs you need to make a potion.

Wolfsbane Potion

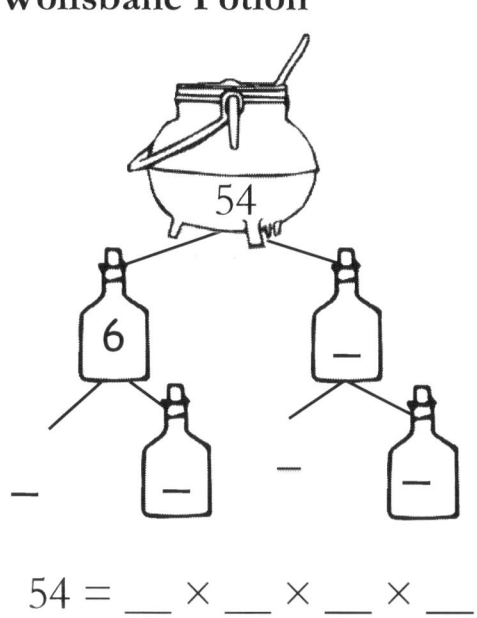

54 = __ × __ × __ × __

Draught of Living Death

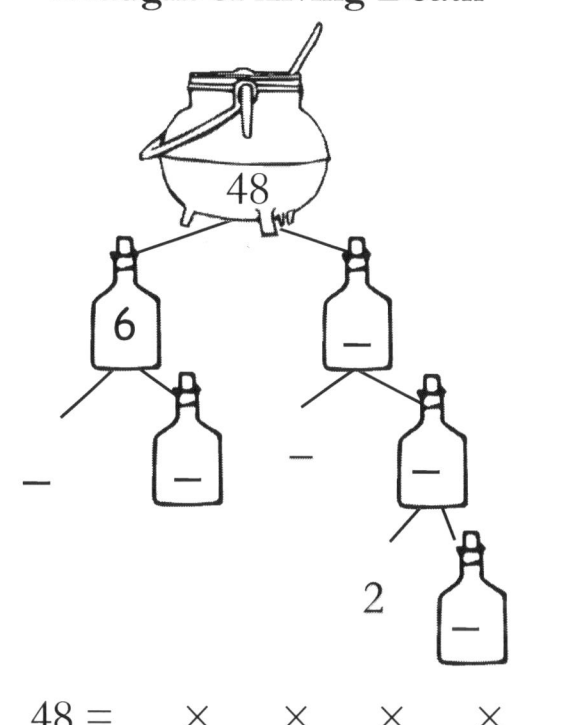

48 = __ × __ × __ × __ × __

Thick Golden Potion

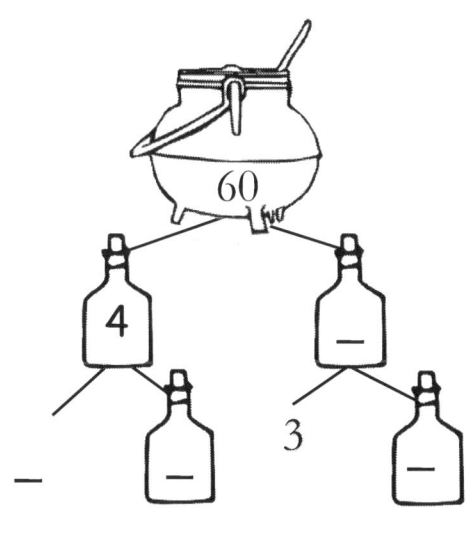

60 = __ × __ × __ × __

Bruise Removal Paste

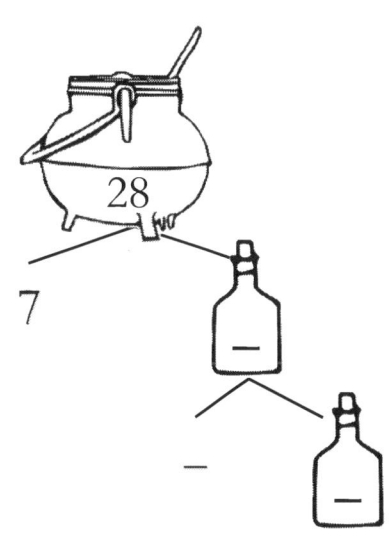

28 = __ × __ × __

1. <u>Multiply</u> and <u>find</u> the value. <u>Indicate</u> the order of operations.

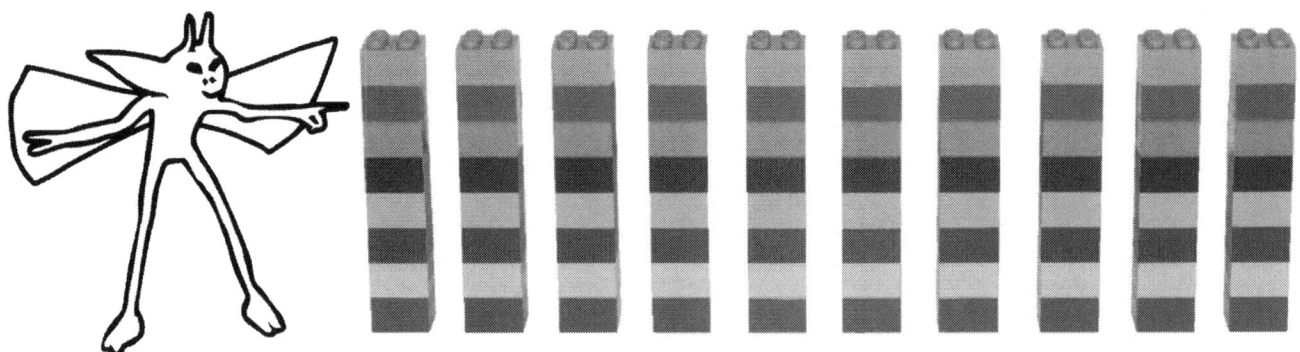

$8 \times 2 = \ldots$ $8 \times 4 = \ldots$ $8 \times 6 = \ldots$ $8 \times 10 = \ldots$

$(8 \times 10) + (8 \times 8) + (8 \times 2) = \ldots$

$5 \times 2 = \ldots$ $5 \times 4 = \ldots$ $5 \times 6 = \ldots$ $5 \times 10 = \ldots$

$(5 \times 8) + (5 \times 10) + (5 \times 2) = \ldots$

$(3 \times 10) + (2 \times 2) + (8 \times 8) =$ _____

$(2 \times 10) + (5 \times 8) + (3 \times 2) =$ _____

$(5 \times 10) + (8 \times 10) + (3 \times 8) =$ _____

$(8 \times 2) - (5 \times 2) + (2 \times 8) =$ _____

Triumph points - …   Rule-breaking points - …

He...didn't dare mention platform nine and three-quarters (see Harry Potter and the Sorcerer's Stone page 76).

Professor Dumbledore offered Dobby ten Galleons a week ... but Dobby beat him down... (see Harry Potter and the Goblet of Fire page 379).

1. How much more would Dobby earn a year if he accepted Professor Dumbledore's offer? Compare with p. 48.

1. <u>Find</u> and <u>circle</u> or <u>cross out</u> the words to find out more about Fawkes.

```
H U H D G Y P R K Z P G N D F S T Y O
T S R E W N E Y R Z O X J B E G X E H
B S A Y A H I H U L Q F M M J A Q K K
O K D F T L G R D F G L A Z N G F R A
D M Y A O F I P E X S L D B A G M U R
N U E X N E L N V T F O Q K D I A T M
L F I K M U L E G O T I E C P N I D U
X T M S M D J I T P X I W V X G O E E
C R Z A K P F N P O O B L H F N U K C
L G G C B C I D L G F W E G H O W C T
O E R M J T A L T D N H E L R I U U N
C R I M S O N I K S F I T R L S A L A
G Y E R I A R Y F E C I R M S E P P I
P X U W H F H C M G M Y C E C D U F L
I B T E P L U F H T I A F F D O L L L
D E C R E P I T L O O K I N G L O A I
P R J K G G G H C J E Y Z H Y J O H R
A L U X F R W R G T J G P D M K T M B
Z G M Y M R O Z I J F X K C L E X U S
```

BRILLIANT

CRIMSON

BURST INTO FLAMES

FAITHFUL PET

HEALING POWERS

DECREPIT-LOOKING

FEATHER

GOLD PLUMAGE

GLITTERING

GAGGING NOISE

SMOLDERING PILE OF ASH

HALF-PLUCKED TURKEY

Triumph points - …

Rule-breaking points - …

1. <u>Multiply</u>.

```
    7         8         7         9         9
×   5     ×   8     ×   6     ×   6     ×   3
_____   _____   _____   _____   _____


    8         4         7         6         4
×   5     ×   9     ×   7     ×   6     ×   4
_____   _____   _____   _____   _____


    4         9         5         2         6
×   5     ×   7     ×   5     ×   9     ×   4
_____   _____   _____   _____   _____
```

2. There are ⬚8 lamp posts along the road of Privet Drive. The lamp posts are ⬚7 ft apart from each other. <u>How far (in inches)</u> is the fifth lamp post from the first one?

Lamp posts	Ft between	Intervals	Inches
___	___	___	___

Answer: _____

1. <u>Answer</u> the question.

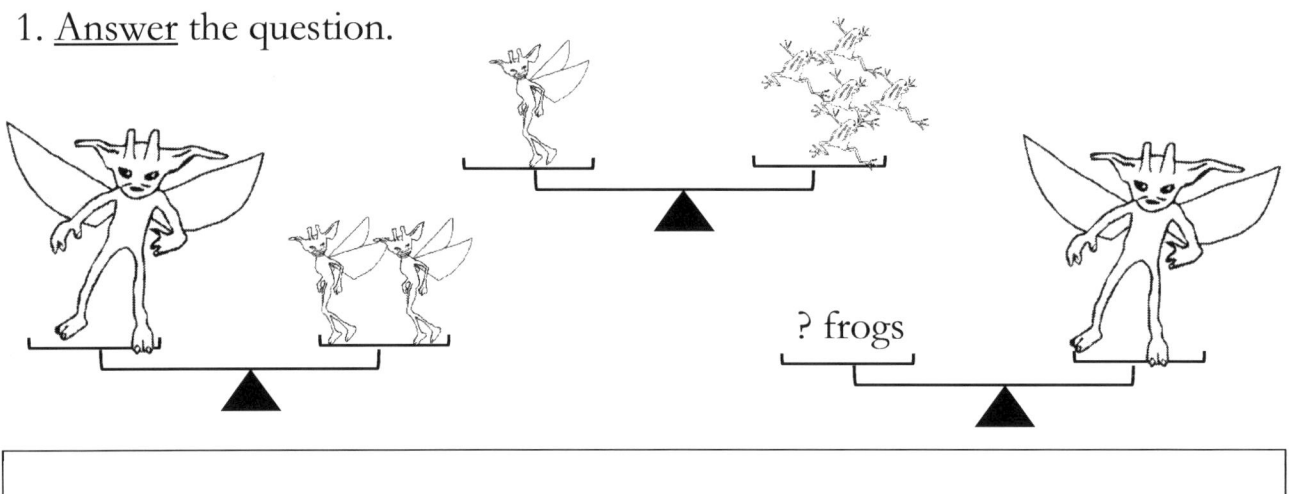

Answer:_____

2. <u>Write in</u> the missing numbers on a potion factor tree. The number shows how many bottles you need, and the bottle shows how many grams of herbs you need to make a potion.

Essence of Anger

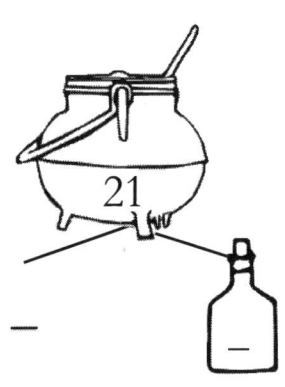

$21 = \underline{} \times \underline{}$

Essence of Dittany

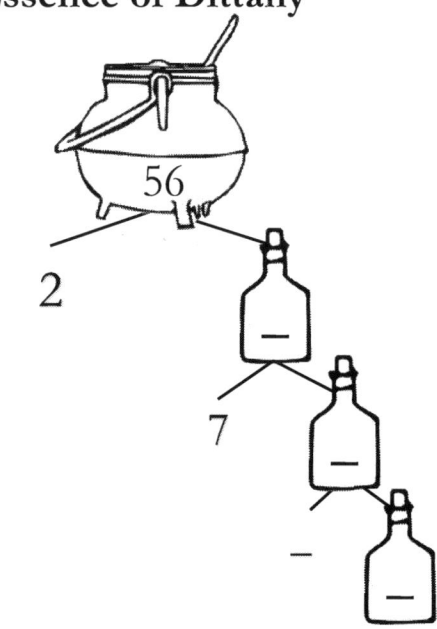

$56 = \underline{} \times \underline{} \times \underline{} \times \underline{}$

Triumph points - … Rule-breaking points - …

1. <u>Divide</u>.

$$2\overline{)4} = 2,\ \text{remainder }0$$

$3\overline{)9}\qquad 4\overline{)12}\qquad 5\overline{)10}\qquad 6\overline{)12}$

$8\overline{)16}\qquad 5\overline{)15}\qquad 7\overline{)14}\qquad 6\overline{)18}\qquad 5\overline{)25}$

$7\overline{)28}\qquad 8\overline{)24}\qquad 9\overline{)27}\qquad 8\overline{)32}\qquad 9\overline{)18}$

2. Harry is thinking of a number. When I multiply 9 to the number, the answer is 36. <u>What</u> is the number?

Answer: _____.

Triumph points - …　　　　　　　　　　Rule-breaking points - …

1. <u>Find</u> the multiplier. Hint: the multiplier tells you how many equal groups you have put together.

_____ times 2 =	4	_____ times 2 is	6
_____ times 2 =	8	_____ times 2 is	10
_____ times 2 =	12	_____ times 2 is	14
_____ times 2 =	16	_____ times 2 is	18
_____ times 3 =	9	_____ times 4 is	16
_____ times 5 =	25	_____ times 6 is	36
_____ times 7 =	49	_____ times 8 is	64
_____ times 9 =	81	_____ times 4 is	12
_____ times 5 =	15	_____ times 6 is	18
_____ times 7 =	21	_____ times 8 is	24
_____ times 9 =	27	_____ times 5 is	20
_____ times 6 =	24	_____ times 7 is	28
_____ times 8 =	32	_____ times 9 is	36
_____ times 6 =	30	_____ times 7 is	35
_____ times 8 =	40	_____ times 9 is	45
_____ times 7 =	42	_____ times 8 is	48
_____ times 9 =	54	_____ times 8 is	56
_____ times 9 =	63	_____ times 9 is	72

Triumph points - …   Rule-breaking points - …

CHAPTER 2

Fun with

Multiplying and Dividing within 1000 without Regrouping

Multiplying and Dividing Tens, Hundreds, and Thousands

Dividing by 10-90s

Learning the Order of Operations

Word Problems

Multiplication and Division Tricks

Word Search

You might belong in Hufflepuff,

Where they are just and loyal,

Those patient Hufflepuffs are true,

And unafraid of toil;

(see *Harry Potter and the Sorcerer's Stone* page 97).

1. <u>Multiply</u>. <u>Use</u> place value to multiply by multiples of 10.

10 × 2 = …	20 × 4 = …	10 × 8 = …	30 × 2 = …
40 × 2 = …	50 × 2 = …	20 × 2 = …	30 × 3 = …
10 × 7 = …	20 × 3 = …	10 × 9 = …	20 × 5 = …

Wow, I wonder what it'd be like to have a difficult life? (see **Harry Potter and the Order of the Phoenix** page 235).

No wonder, a difficult life is a life with MATH.

Are you kidding me? That's the easiest problem we've ever done!

Look, 10 × 2 is 1 ten times two or ten 2's: 1 × 10 × 2.

Hint: here is a trick. You multiply 1 by 2 or 1 times 2.

Then, write 0 (zero) in one's place: 10 × 2 = 20.

Let's solve one more problem: 40 × 2 is 4 tens times 2: 4 × 10 × 2.

Multiply 4 times 2 and write 0 (zero) in one's place: 40 × 2 = 80.

See, Scaredy-cat!

I see… If I have 50 × 2, it means 5 tens times 2, right?

5 times 2 is 10, and I write 0 (zero) in one's place:

50 × 2 = 5 × 10 × 2 = 100.

And I'm NOT a Scaredy-cat!

Triumph points - …   Rule-breaking points - …

...I want to fix that in my memory forever, (see **Harry Potter and the Goblet of Fire** page 207).

No, I don't want to fix this trick in my memory forever –

multiply here, write zero there!

I'm going to keep going until I succeed – or I die (see **Harry Potter and the Deathly Hallows** page 569).

No drama, guys, we've gotta make it out alive! You multiply each number starting from ones, then tens, then hundreds, and so on:

4 times 0 equals 0, and you write 0 in one's place:

$20 \times 4 = \ldots 0.$

Then, 4 times 2 equals 8 and you write 8 in ten's place:

$20 \times 4 = 80.$

Hint: use the small number for the multiplier!

WRONG:
```
    4
×  2 0
─────
   8 0
```

Triumph points - … Rule-breaking points - …

1. <u>Multiply</u>.

$4 \times 20 = \ldots$ $2 \times 10 = \ldots$ $3 \times 30 = \ldots$ $5 \times 20 = \ldots$

$3 \times 20 = \ldots$ $2 \times 40 = \ldots$ $6 \times 10 = \ldots$ $2 \times 20 = \ldots$

$9 \times 10 = \ldots$ $5 \times 10 = \ldots$ $3 \times 10 = \ldots$ $8 \times 10 = \ldots$

$\ldots$ / 4×8 $\ldots$ / 5×6 $\ldots$ / 8×3 $\ldots$ / 5×5 $\ldots$ / 7×7

$\ldots$ / 8×8 $\ldots$ / 8×6 $\ldots$ / 4×7 $\ldots$ / 9×9 $\ldots$ / 9×3

2. <u>Multiply</u> and <u>find</u> the product. The first factor is ⟦two⟧ bronze Knuts. The first one is done for you.

$50 \times \ldots = \ldots$

$40 \times 2 = 80$

$30 \times \ldots = \ldots$

$80 \times \ldots = \ldots$ $90 \times \ldots = \ldots$ $70 \times \ldots = \ldots$

$20 \times \ldots = \ldots$ $60 \times \ldots = \ldots$

Triumph points - …   Rule-breaking points - …

"Hope to see you in Hufflepuff! ... My old House, you know (see *Harry Potter and the Sorcerer's Stone* page 96).

Well, I like this House, too."

1. There are 27 pixies in all. I divided the pixies into groups of 3. <u>How many pixies</u> are big?

1. <u>Find</u> and <u>circle</u> or <u>cross out</u> the words to find out more about Malfoy family.

```
S L D D H C G T E R R E F J Y N S
T K G R K R G N O A R W D J O R L
M Q H A N U K C I X K E V O F O Y
L C D G T C W A H R A F Q B L H T
H X Z O Y I Q B E D T R V O A T H
G E G N L A B U P B K S V F M W E
O S S I E T J A I U K M T T O A R
D M E Y K U N G O D N C S R C H I
B R O O M S T I C K D I U E A R N
O Q G E T C N R K S L I C B R E X
S Y Y A I U Z G E L N D T O D E H
U Z R N U R D E E P O Q W C R O I
B E O T L S K U J S V Q T T H N W
M I H O W E D A Z K A B A N H A W
I O R S R Y O F L A M S U I C U L
N P R E E N O I T O P F J Z X H K
C B T R U W V U K J P K J I D O T
```

DRAGON	SEEKER	QUIDDITCH
HEARTSTRING	SLYTHERIN	FERRET
CRUCIATUS CURSE	POTIONEER	BUCKBEAK
DEADPAN STARE	AZKABAN	LUCIUS MALFOY
DRACO MALFOY	HAWTHORN	NIMBUS
UNICORN	BROOMSTICK	DUELLIST

Triumph points - …   Rule-breaking points - …

1. <u>Answer</u> the question.

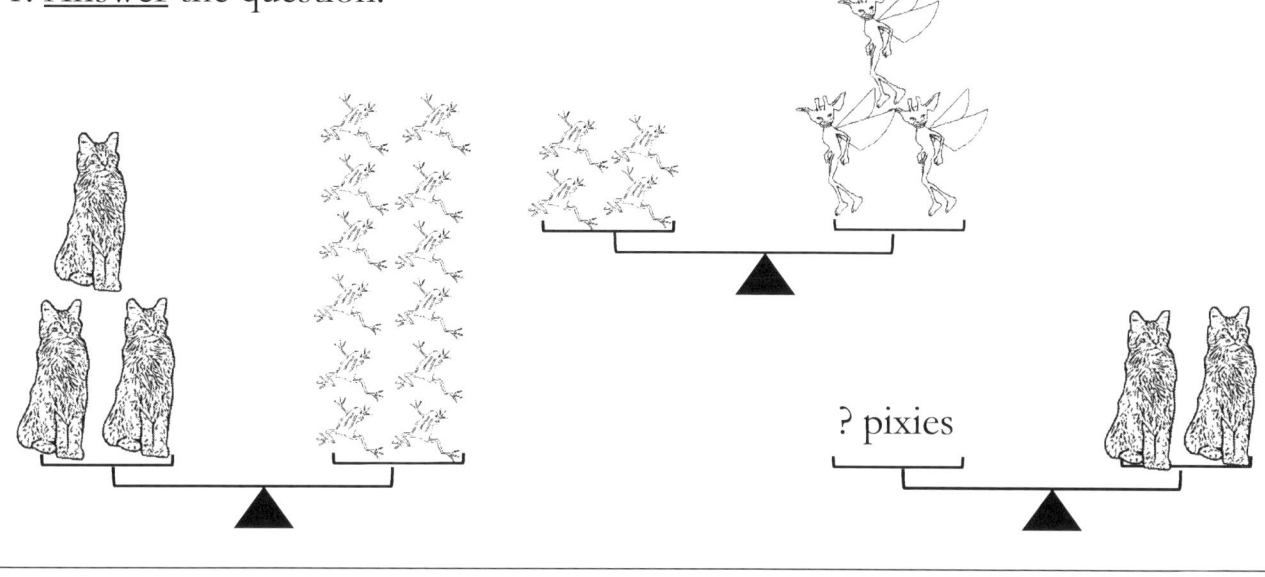

Answer: _____

2. <u>Multiply</u>.

```
   20        50        80        60        60
 ×  5      ×  8      ×  2      ×  4      ×  9
  100       400       160       240       540
```

```
   80        40        70        80        80
 ×  4      ×  7      ×  3      ×  7      ×  6
  320       280       210       560       480
```

```
   60        90        70        50        60
 ×  5      ×  8      ×  5      ×  6      ×  8
  300       720       350       300       480
```

1. Divide.

$$2\overline{)160} = 80 \qquad 3\overline{)300} = 100 \qquad 4\overline{)400} = 100 \qquad 5\overline{)500} = 100$$

$$8\overline{)800} = 100 \qquad 5\overline{)200} = 40 \qquad 7\overline{)700} = 100 \qquad 6\overline{)120} = 20$$

$$7\overline{)350} = 50 \qquad 8\overline{)320} = 40 \qquad 9\overline{)450} = 50 \qquad 8\overline{)240} = 30$$

2. Write the missing one-digit numbers where the quotient should be smaller than the divisor.

$34 \div \underline{} = \underline{}\ r\ 2 \qquad 22 \div \underline{} = \underline{}\ r\ 1 \qquad 57 \div \underline{} = \underline{}\ r\ 1$

$20 \div \underline{} = \underline{}\ r\ 2 \qquad 38 \div \underline{} = \underline{}\ r\ 3 \qquad 61 \div \underline{} = \underline{}\ r\ 5$

$75 \div \underline{} = \underline{}\ r\ 3 \qquad 49 \div \underline{} = \underline{}\ r\ 1 \qquad 82 \div \underline{} = \underline{}\ r\ 2$

Triumph points - …

Rule-breaking points - …

1. <u>Write in</u> the missing numbers on a potion factor tree. The number shows how many bottles you need, and the bottle shows how many grams of herbs you need to make a potion.

Love Potion

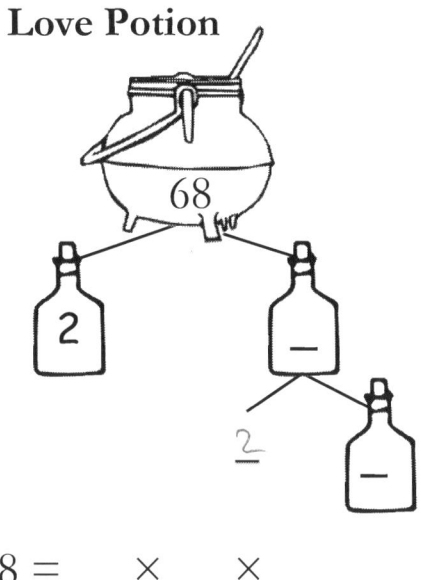

68 = __ × __ × __

Love Potion Antidote

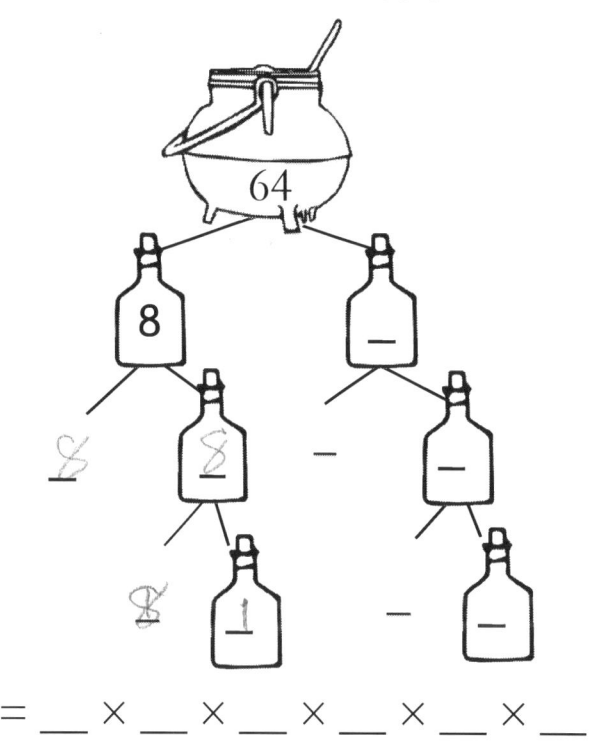

48 = __ × __ × __ × __ × __ × __

Silence Potion

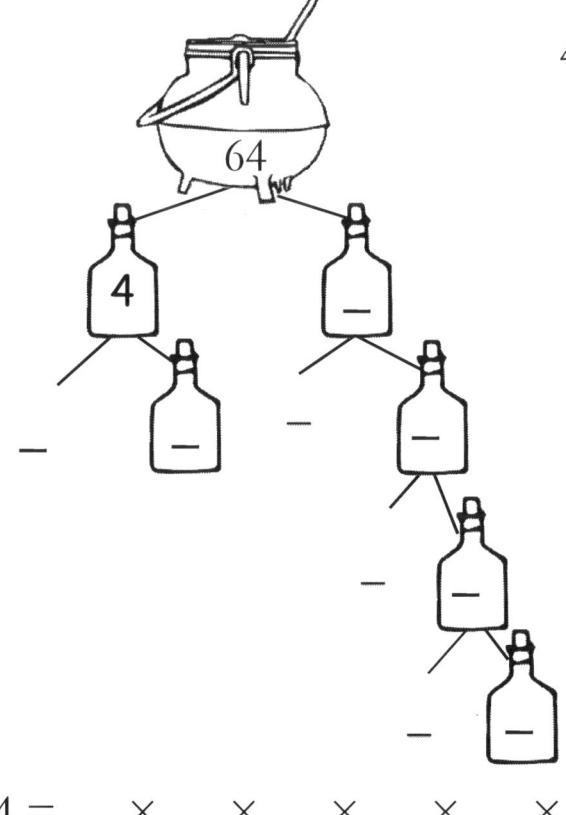

64 = __ × __ × __ × __ × __ × __

Truth Potion Antidote

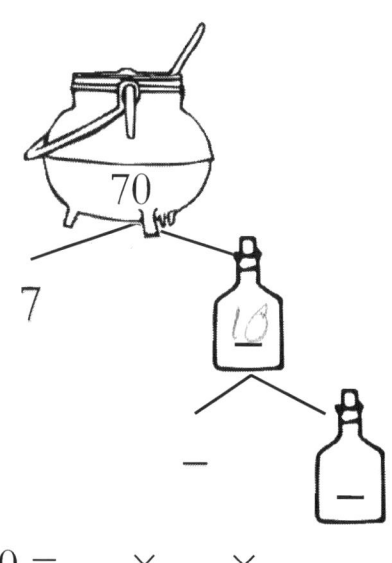

70 = __ × __ × __

Triumph points - …

Rule-breaking points - …

Brainers, I will do division. I like tricks, may I start with the tricks?

When we divide 60 by 2, it means we divide 6 tens by 2:

$(6 \times 10) \div 2$.

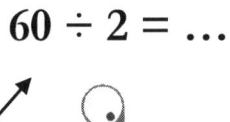

$60 \div 2 = \ldots$

60 divided by 2

Or I can rewrite the problem as: $(6 \div 2) \times 10$.
Step 1: divide 6 by 2. The answer is 3. ⟶ $(6 \div 2) \times 10 = 3\ldots$

Step 2: multiply 3 by 10. The answer is 30. Or write 0 in one's place. ⟶ $(6 \div 2) \times 10 = 30$

I can also write the problem as:

$6 \times (10 \div 2) = 6 \times 5 = 30$.

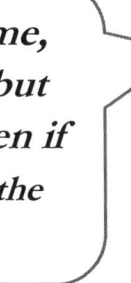

A lot of wizards think it's a waste of time, knowing this sort of Muggle trick, … but we feel they're skills worth learning, even if they are a bit slow (see **Harry Potter and the Chamber of Secrets** page 26).

So, divide 40 by 2.

Step 1: 4 divided by 2 equals 2. Write 2 in ten's place.

Multiply to check your answer: $2 \times 2 = 4$.

$40 \div 2 = 2 \ldots$

$$\begin{array}{r} 2 \\ 2\overline{)40} \end{array}$$

Step 2: 0 divided by 2 is 0.

Write 0 in one's place:

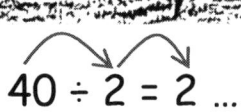

$40 \div 2 = 2\,0$

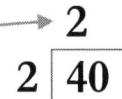

$$\begin{array}{r} 20 \\ 2\overline{)40} \end{array}$$

Triumph points - …

Rule-breaking points - …

1. <u>Divide</u>.

40 ÷ 2 = …	80 ÷ 4 = …	60 ÷ 2 = …	90 ÷ 3 = …
70 ÷ 7 = …	80 ÷ 2 = …	60 ÷ 3 = …	80 ÷ 8 = …
50 ÷ 5 = …	90 ÷ 9 = …	30 ÷ 3 = …	100 ÷ 5 = …
40 ÷ 4 = …	80 ÷ 8 = …	60 ÷ 6 = …	100 ÷ 2 = …

10	30	60	70	20
2 × …	2 × …	2 × …	2 × …	2 × …

90	80	100	40	50
2 × …	2 × …	2 × …	2 × …	2 × …

2. <u>Divide</u> and <u>find</u> the quotient. The divisor is ⬚two⬚ silver Sickles. The first task is done for you.

60 ÷ … = …

100 ÷ … = …

180 ÷ 2 = 90

40 ÷ … = …

140 ÷ … = …

80 ÷ … = …

160 ÷ … = …   120 ÷ … = …

Triumph points - …   Rule-breaking points - …

1. <u>Multiply</u>. <u>Use</u> five multiplication ways for each problem. The first one is done for you.

12 × 4 = 4 × 3 × 2 × 2 = 2 × 6 × 4 × 1 =
= 24 × 2 = 6 × 8 = 2 × 2 × 3 × 2 × 2 =
= 2 × 2 × 3 × 2 × 2 = 48

14 × 6 = _____

_____

16 × 4 = _____

_____

15 × 6 = _____

_____

2. <u>Help</u> me get the Triwizard Cup.

Triumph points - …

Rule-breaking points - …

1. Harry spent ⬚33⬚ silver Sickles on Chocolate Frogs, Pumpkin Pasties, and Licorice Wands. Chocolate Frogs cost ⬚6⬚ silver Sickles, and Pumpkin Pasties cost ⬚four times⬚ as much as Chocolate Frogs. <u>How much more or less</u> did he spend on Chocolate Frogs than on Licorice Wands?

Spent altogether	Frogs	Pasties	Wands	Frogs > or < Wands
___	___	?___	?___	?___

Answer: _____

2. <u>Write in</u> the missing numbers on a potion factor tree. The number shows how many bottles you need, and the bottle shows how many grams of herbs you need to make a potion.

Essence of Happiness

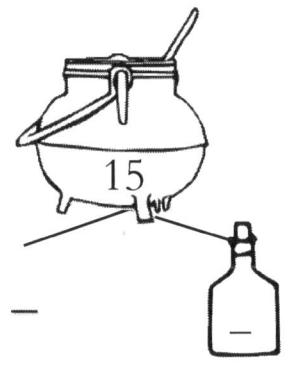

15 = __ × __

Essence of Sadness

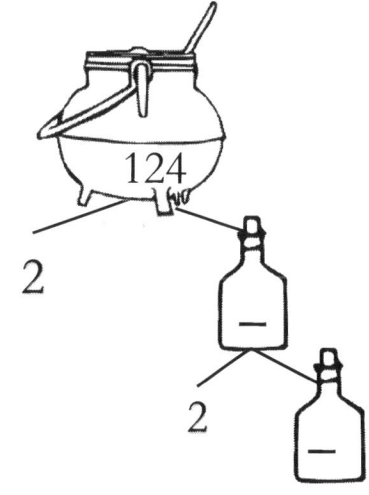

124 = __ × __ × __

Triumph points - …

Rule-breaking points - …

1. <u>Multiply</u>.

```
   …            …            …            …            …
  ╱ ╲          ╱ ╲          ╱ ╲          ╱ ╲          ╱ ╲
 2 × 9        3 × 7        6 × 3        5 × 8        2 × 6

   …            …            …            …            …
  ╱ ╲          ╱ ╲          ╱ ╲          ╱ ╲          ╱ ╲
 7 × 5        7 × 8        6 × 7        8 × 8        4 × 9
```

2. <u>Divide</u>.

```
  20           40           60           80          100
  ╱ ╲          ╱ ╲          ╱ ╲          ╱ ╲          ╱ ╲
 2 × …        2 × …        2 × …        2 × …        2 × …

  30           90           60           50          100
  ╱ ╲          ╱ ╲          ╱ ╲          ╱ ╲          ╱ ╲
 3 × …        3 × …        3 × …        5 × …        5 × …
```

3. <u>Multiply</u> and <u>find</u> the product. The first factor is three bronze Knuts. The first task is done for you.

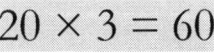

20 × 3 = 60

90 × … = …

50 × … = …

30 × … = …

70 × … = …

60 × … = …

80 × … = …

40 × … = …

Triumph points - …

Rule-breaking points - …

 Your home is My Dream School!.. Are there any scholarships for Muggles?!

 Don't know. Try to solve the problem.

1. There are 32 pixies in all. I divided the pixies into groups of 3. How many pixies are small?

Triumph points - ... Rule-breaking points - ...

"I wonder, how do you divide by 10 or 100 or even 1000?"

Is he – a bit mad?
(see **Harry Potter and the Sorcerer's Stone** page 101).

"I'm not so sure… Nevermind, look, I divide 80 by 20:

Hint! Step 1: Both numbers have 0's, so, cross out these 0's:

Step 2: divide the digits:

Step 3: write the answer:"

$80 \div 20 = \ldots$

$8\cancel{0} \div 2\cancel{0} = \ldots$

$8\cancel{0} \div 2\cancel{0} = 8 \div 2 = \ldots$

$8\cancel{0} \div 2\cancel{0} = 8 \div 2 = 4$

"Just keep in mind that you need to cross out the SAME number of 0's in both the dividend and the divisor! Let me show you:

$600 \div 30 =$ Cross out only one 0 (zero) $= 60\cancel{0} \div 3\cancel{0} = 60 \div 3 = 20$."

… an' don' ask me questions just now, I think I'm gonna be sick (see **Harry Potter and the Sorcerer's Stone** page 65).

"Show me another way!"

Triumph points - …

Rule-breaking points - …

I hope the Muggles didn't give you a hard time (see *Harry Potter and the Prisoner of Azkaban* page 9).

I will help you! Look, I divide 160 by 20.

$160 \div 20 = \ldots$

How many 20's are in 160?

The closest answer is eight 20's:

Check your answer:

Step 1: $8 \times 0 = 0$

Step 2: $8 \times 2 = 16$

Step 3: $160 - 160 = 0$

```
      8      Step1
 20 | 160
        0    8×0
```

```
      8      Step2
 20 | 160
    -160     8×2
       0
```

And you can always **multiply to check your answer**:

$8 \times 20 = 8 \times 2 \times 10 = 16 \times 10 = 160$.

1. <u>Divide</u>. <u>Cross out</u> 0's in both the dividend and the divisor.

$1\cancel{5}\cancel{0} \div 1\cancel{0} = 15$ $300 \div 30 = \ldots$ $600 \div 20 = \ldots$

$840 \div 10 = \ldots$ $80 \div 40 = \ldots$ $90 \div 30 = \ldots$

$550 \div 10 = \ldots$ $120 \div 30 = \ldots$ $140 \div 70 = \ldots$

$560 \div 80 = \ldots$ $810 \div 90 = \ldots$ $360 \div 40 = \ldots$

Triumph points - …   Rule-breaking points - …

1. <u>Answer</u> the question.

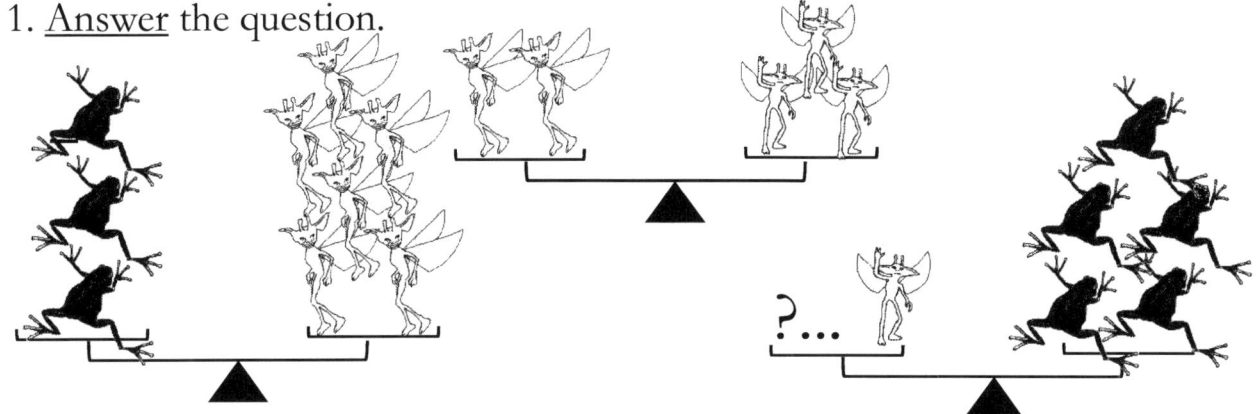

Answer:_____

2. <u>Write in</u> the missing numbers on a potion factor tree. The number shows how many bottles you need, and the bottle shows how many grams of herbs you need to make a potion.

Essence of Laziness

49 = __ × __

Essence of Naughtiness

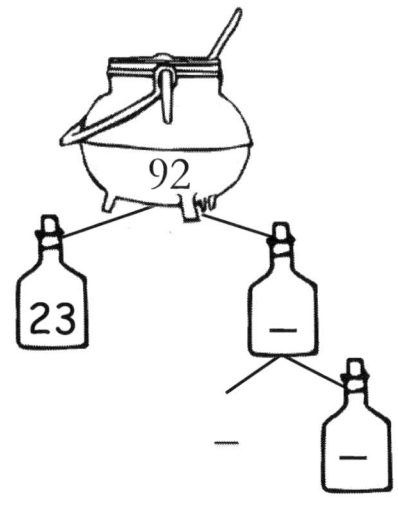

94 = __ × __ × __

Triumph points - …

Rule-breaking points - …

1. <u>Multiply</u>.

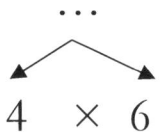   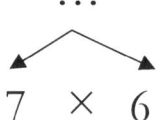   ...   ...   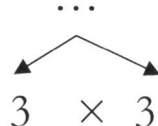

4 × 6   7 × 6   8 × 9   5 × 7   3 × 3

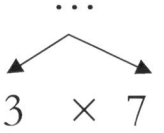

3 × 7   5 × 6   4 × 3   8 × 4   9 × 5

2. <u>Divide</u>. <u>Cross out</u> [0's].

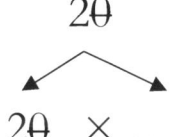

2̶0̶   40   60   80   100

2̶0̶ × …   20 × …   20 × …   20 × …   20 × …

30   90   60   80   100

30 × …   30 × …   30 × …   40 × …   50 × …

3. <u>Divide</u> and <u>find</u> the quotient. The divisor is [three] bronze Knuts. The first task is done for you.

90 ÷ … = …

270 ÷ … = …

60 ÷ … = …

180 ÷ 3 = 90

240 ÷ … = …

120 ÷ … = …

210 ÷ … = …

150 ÷ … = …

Triumph points - …   Rule-breaking points - …

1. <u>Find</u> and <u>circle</u> or <u>cross out</u> the words to find out more Hogwarts' Professors.

```
F Y G Z N F Y K L R K V K K L Q P S S G
N F C S M R E S E O I C C P Y B O G N I
F C V N E Y O M S B Y I S R R D M A N L
Z N N N Q V U H H C W E M L O P O Y I D
T G X K Q S E V G T C E G L N N N O B E
J H Z E L U X R I U V G O E R U A Z R R
G R C U J N I L U D L R C C E S S F O O
W X P O R R F R H S E S I S Q M P X S Y
V I A E M S L O I S S V E B O X R A S L
N B Z J U H F S U N B N C C Q G O Z E O
U J T I R J W M W Y U A A X A E U W F C
Z W L S C A B Z X G Q S A P M R T X O K
W I D Y I R G C B B Y Q Q U E D O I R H
F B F A I N E Z P M L V J U F G N H P A
C C U D S K R S M Z O I U R I U D K C R
A G G B A R T Y C R O U C H J R W J K T
Y E N W A L E R T L L I B Y S N R X E P
M I N E R V A M C G O N A G A L L E K Y
Y U Q B K G S V M B S Y A P P J S L P
G P J S A O D Q P R G C F I E Y D I A L
```

REMUS LUPIN   HORACE SLUGHORN

PROFESSOR BINNS   BARTY CROUCH JR.

MINERVA MCGONAGALL   SEVERUS SNAPE

DOLORES UMBRIDGE   SYBILL TRELAWNEY

QUIRINUS QUIRRELL   POMONA SPROUT

FILIUS FLITWICK   GILDEROY LOCKHART

Triumph points - …   Rule-breaking points - …

1. <u>Multiply</u>.

```
   70        90        40        70        50
×   5     ×   8     ×   2     ×   4     ×   9
_____     _____     _____     _____     _____


   30        70        80        90        60
×   4     ×   7     ×   3     ×   7     ×   6
_____     _____     _____     _____     _____


   40        70        80        30        20
×   5     ×   8     ×   5     ×   6     ×   8
_____     _____     _____     _____     _____
```

2. Mr. Dursley paid $24 in all for 6 tickets to the zoo. <u>How many three-dollar tickets</u> did he buy? <u>How many five-dollar tickets</u> did he buy?

Paid in all	Tickets in all	Three-dollar tickets	Five-dollar tickets
___	___	? ___	? ___

Answer: _____

1. <u>Divide</u>.

$2\overline{)180}$ $3\overline{)210}$ $4\overline{)240}$ $5\overline{)350}$

$8\overline{)560}$ $5\overline{)400}$ $7\overline{)490}$ $6\overline{)300}$

$7\overline{)490}$ $8\overline{)640}$ $9\overline{)810}$ $8\overline{)160}$

2. <u>Write</u> the missing one-digit numbers where the remainder is the greatest.

4__ ÷ 7 = 5 r __   4__ ÷ 6 = 6 r __   2__ ÷ 3 = 8 r __

3__ ÷ 4 = 9 r __   1__ ÷ 2 = 6 r __   4__ ÷ 8 = 5 r __

7__ ÷ 9 = 7 r __   5__ ÷ 6 = 8 r __   2__ ÷ 5 = 4 r __

Triumph points - …   Rule-breaking points - …

1. <u>Write in</u> the missing numbers on a potion factor tree. The number shows how many bottles you need, and the bottle shows how many grams of herbs you need to make a potion.

Wideeye Potion

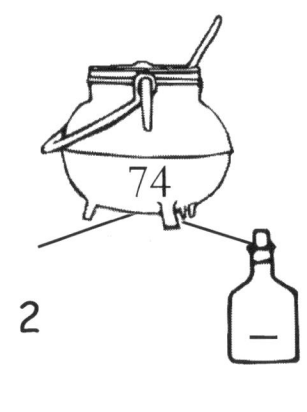

$74 = \_\_ \times \_\_$

Beauty-making Potion

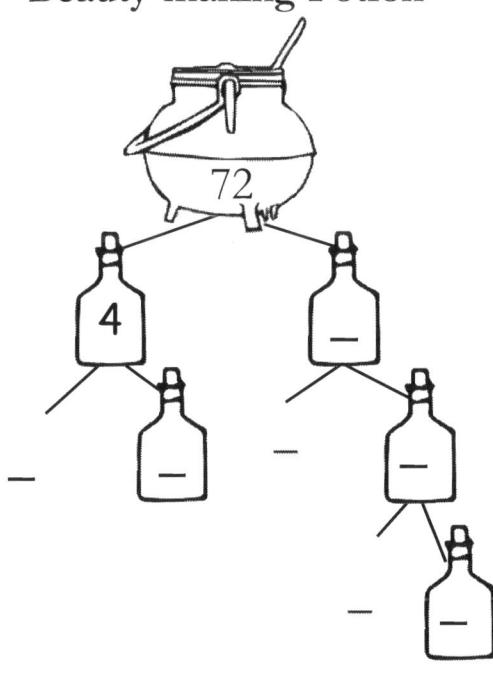

$72 = \_\_ \times \_\_ \times \_\_ \times \_\_ \times \_\_$

Beauty-making Potion Antidote

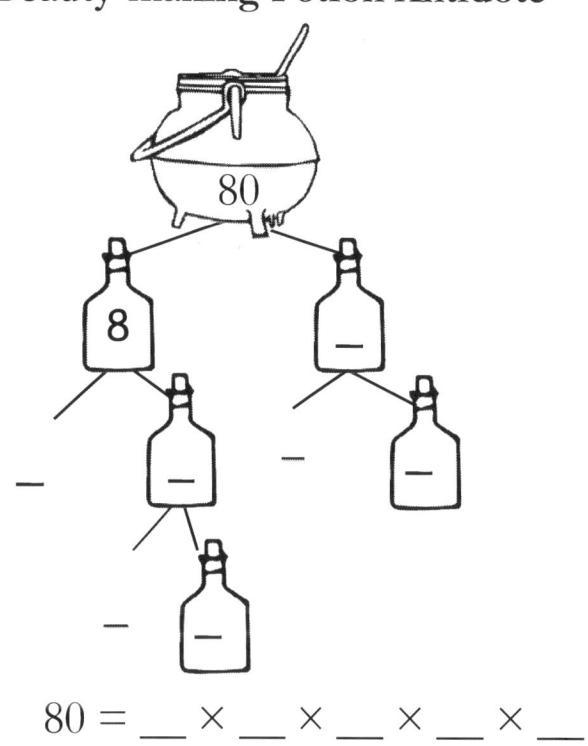

$80 = \_\_ \times \_\_ \times \_\_ \times \_\_ \times \_\_$

Burning Potion

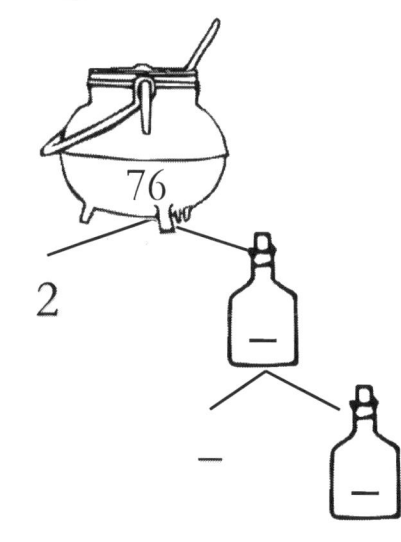

$76 = \_\_ \times \_\_ \times \_\_$

Triumph points - …   Rule-breaking points - …

1. <u>Answer</u> the questions.

<u>Continue</u> a series of numbers (multiplication).

4, 8, 24, 48, _____, _____

Think, Brainer, think!

Don't let the Muggles get you down! (see **Harry Potter and the Prisoner of Azkaban** page 10).

2. <u>Find</u> the 🤭.

🤭 ÷ (36 − 27) = 9   (34 + 29) ÷ 🤭 = 7

_____   _____

🤭 = _____   🤭 = _____

3. <u>Multiply</u> and <u>find</u> the product. The first factor is five bronze Knuts. The first task is done for you.

20 × 5 = 100

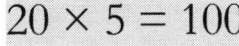

60 × … = …   30 × … = …

  90 × … = …   80 × … = …

50 × … = …   70 × … = …   40 × … = …

Triumph points - …   Rule-breaking points - …

1. <u>Multiply</u> or <u>divide</u>.

It'll be a lot less hassle if you can just knock Malfoy off his broom tomorrow (see *Harry Potter and the Chamber of Secrets* page 166).

He's right, you know? I wish I could knock my math off, I'm so mad.

100 × 3 =

1000 ÷ 10 =

200 × 5 =

200 × 2 =

600 ÷ 6 =

320 ÷ 4 =

10 × 100 =

490 ÷ 7 =

30 × 7 =

60 × 3 =

800 ÷ 4 =

350 ÷ 5 =

40 × 8 =

90 ÷ 9 =

70 × 8 =

70 × 4 =

540 ÷ 6 =

1000 ÷ 5 =

Oh, my! I'm lost in the clouds! Not again! Watch out! Dementors!

810 ÷ 9 =

50 × 8 =

420 ÷ 6 =

360 ÷ 4 =

50 ÷ 5 =

10 × 5 =

70 × 9 =

50 × 5 =

270 ÷ 3 =

40 × 6 =

Triumph points - …

Rule-breaking points - …

1. How many 10's are in the highlighted numbers?

620:    $620 \div 10 = 62$ 890:   ... ÷ ... = ...

1010:   ... ÷ ... = ...   270:   ... ÷ ... = ...

550:    ... ÷ ... = ...   730:   ... ÷ ... = ...

380:    ... ÷ ... = ...   960:   ... ÷ ... = ...

2. Multiply and divide.

```
   ...            ...            ...            ...            ...
  /   \          /   \          /   \          /   \          /   \
20 × 5        30 × 7        50 × 9        40 × 6        20 × 7

   ...            ...            ...            ...            ...
  /   \          /   \          /   \          /   \          /   \
90 × 3        70 × 4        50 × 8        30 × 5        20 × 9

  350            560            810            490            640
  /   \          /   \          /   \          /   \          /   \
7 × ...       8 × ...       9 × ...       7 × ...       8 × ...

  630            400            720            280            540
  /   \          /   \          /   \          /   \          /   \
7 × ...       8 × ...       9 × ...       7 × ...       6 × ...
```

Triumph points - ...   Rule-breaking points - ...

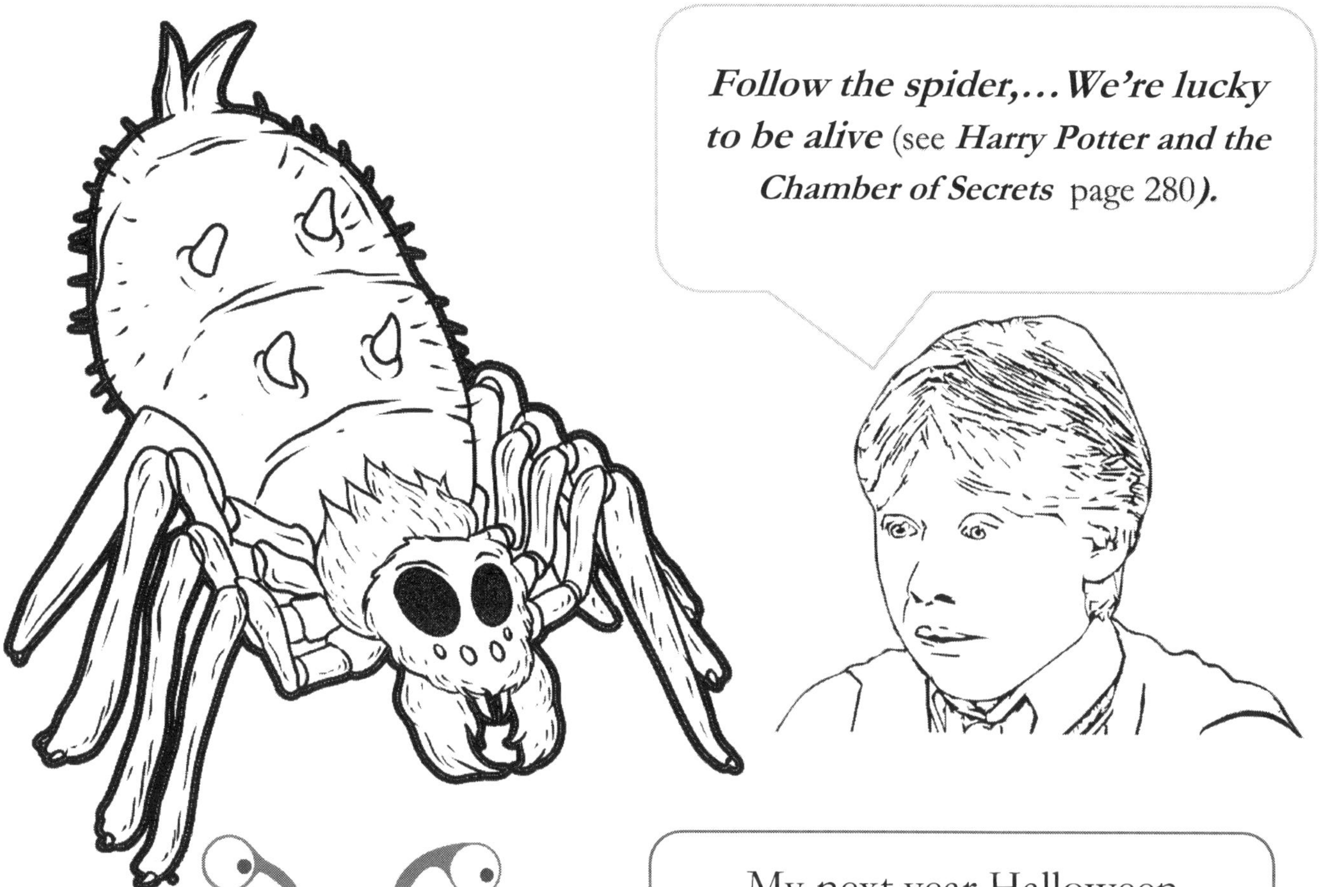

Follow the spider,... We're lucky to be alive (see ***Harry Potter and the Chamber of Secrets*** page 280*).*

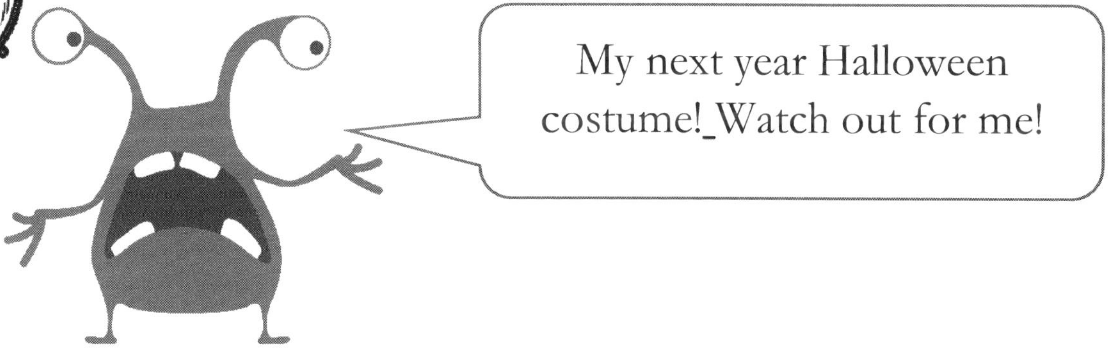

My next year Halloween costume!_Watch out for me!

1. Some towers with bricks are hidden behind the question mark (?). <u>How many towers</u> are there?

? ...

Triumph points - ...

Rule-breaking points - ...

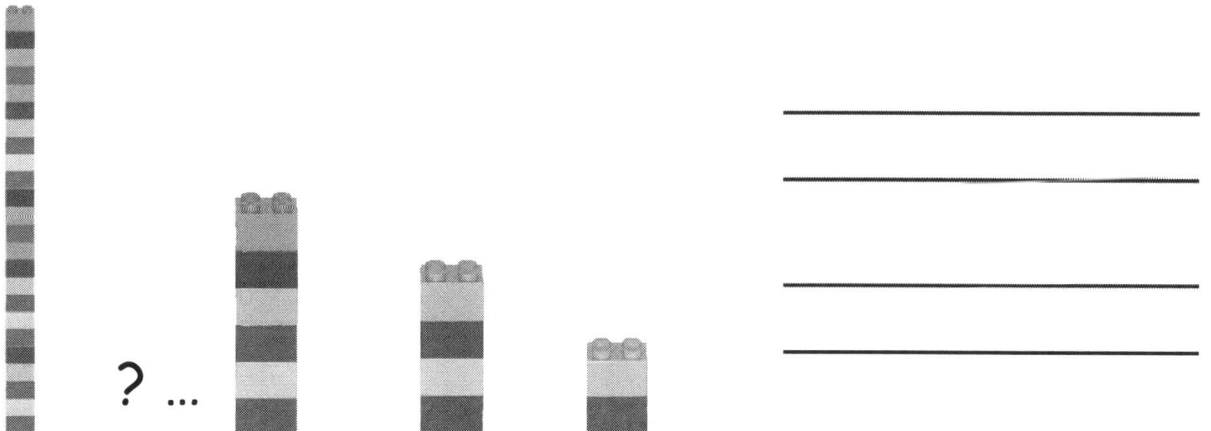

1. <u>Draw</u> the frogsss for ssstage sss3.

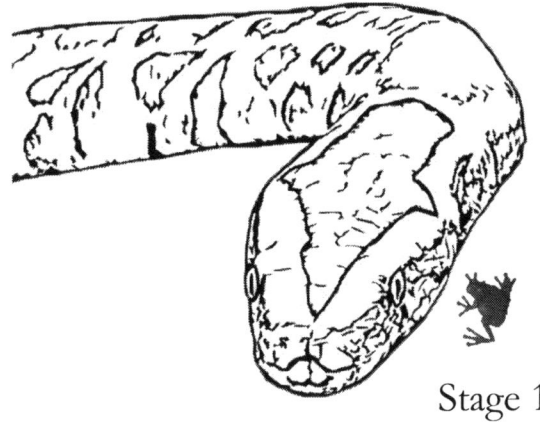

Stage 1 Stage 2 Stage 3

2. <u>Find</u> the 🤪.

(🤪 ÷ 8) + 4 = 9   (50 - 45) × 🤪 = 35

_____   _____

🤪 = _____   🤪 = _____

3. <u>Divide</u> and <u>find</u> the quotient. The divisor is five bronze Knuts. The first task is done for you.

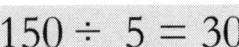 150 ÷ 5 = 30   200 ÷ … = …

250 ÷ … = …   450 ÷ … = …

   350 ÷ … = …

500 ÷ … = …   300 ÷ … = …   400 ÷ … = …

Triumph points - …   Rule-breaking points - …

1. <u>Answer</u> the question.

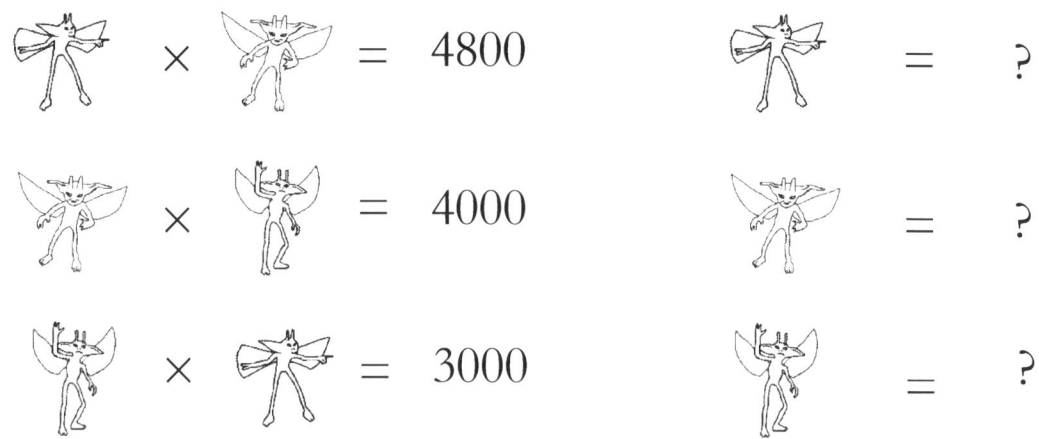

2. There are 13 flying cars and flying bikes in Hogwarts' parking lot. There are a total of 40 wheels. <u>How many flying cars</u> are in the Hogwarts' parking lot? <u>How many flying bikes</u> are in the parking lot?

Cars & bikes	Wheels in all	Cars	Bikes
___	___	? ___	? ___

Answer: _____

Triumph points - … Rule-breaking points - …

1. Answer the questions.

1. How many 20's are in the highlighted numbers?

800: … ÷ … = …   680: … ÷ … = …

1000: … ÷ … = …   440: … ÷ … = …

How many 40's are in the highlighted numbers?

800: … ÷ … = …   480: … ÷ … = …

How many 50's are in the highlighted numbers?

550: … ÷ … = …   1500: … ÷ … = …

2. Multiply and find the product. The first factor is four bronze Knuts.

50 × … = …

40 × … = …

20 × … = …   70 × … = …

4 × 90 = 360

60 × … = …

80 × … = …

30 × … = …

3. Divide and find the quotient. The divisor is four bronze Knuts.

Training for the ballet … ? (see *Harry Potter and the Chamber of Secrets* page 171).

80 ÷ … = …

360 ÷ … = …

320 ÷ … = …

200 ÷ … = …

240 ÷ … = …   160 ÷ 4 = 40   280 ÷ … = …   120 ÷ … = …

Triumph points - …   Rule-breaking points - …

1. <u>Multiply</u>. <u>Use</u> five multiplication ways for each problem (decompose the factors). <u>Manipulate</u> the factors!

15 × 4 = _____

_____

18 × 6 = _____

_____

20 × 4 = _____

_____

25 × 4 = _____

_____

2. <u>Help</u> me get the Triwizard Cup.

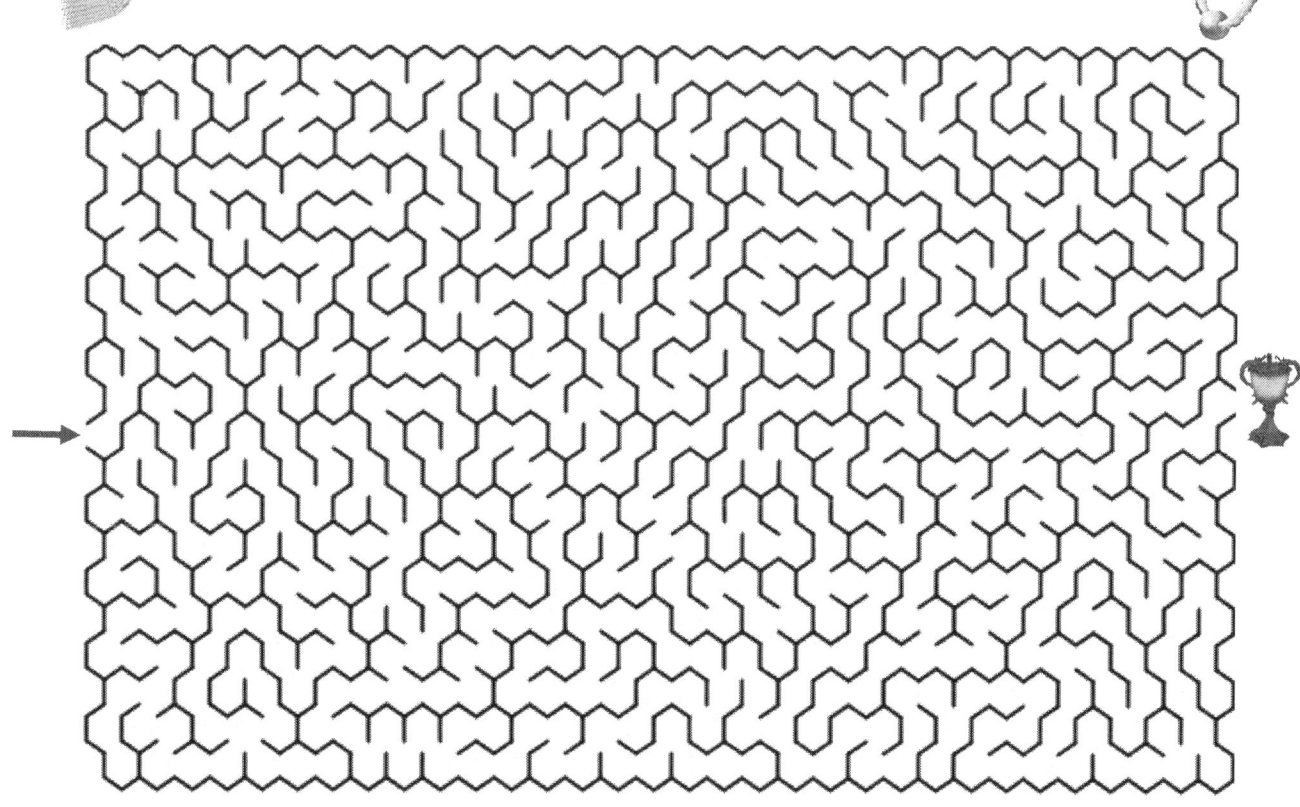

Triumph points - … Rule-breaking points - …

1. <u>Evaluate</u> each expression. First, start with the parentheses (brackets), then, division or multiplication OR addition or subtraction from left to right! <u>Indicate</u> the order of operation.

$(54 \div 9)^1 +^4 (18 \div 9)^2 -^5 (18 \div 6)^3 = \ldots$ $(81 \div 9)^1 \div^3 (54 \div 6)^2 = \ldots$

$(27 \div 9) + (36 \div 6) - (36 \div 9) = \ldots$ $(48 \div 6) \div (72 \div 9) = \ldots$

$(63 \div 9) + (42 \div 6) - (81 \div 9) = \ldots$ $(30 \div 6) \times (45 \div 9) = \ldots$

$(45 \div 9) + (72 \div 8) - (27 \div 9) = \ldots$ $(72 \div 9) \div (32 \div 8) = \ldots$

$(18 \div 9) + (63 \div 9) - (56 \div 8) = \ldots$ $(81 \div 9) \div (24 \div 8) = \ldots$

$(16 \div 8) + (36 \div 9) - (45 \div 9) = \ldots$ $(48 \div 8) \times (54 \div 9) = \ldots$

$(21 \div 3) + (35 \div 7) - (49 \div 7) = \ldots$ $(63 \div 9) \times (9 \div 3) = \ldots$

$(42 \div 7) + (27 \div 3) - (14 \div 7) = \ldots$ $(28 \div 7) \div (6 \div 3) = \ldots$

$(21 \div 7) + (24 \div 3) - (35 \div 7) = \ldots$ $(56 \div 7) \times (12 \div 3) = \ldots$

$(28 \div 7) + (63 \div 7) - (36 \div 4) = \ldots$ $(36 \div 4) \div (63 \div 7) = \ldots$

$(35 \div 7) + (16 \div 4) - (21 \div 7) = \ldots$ $(56 \div 7) \times (24 \div 4) = \ldots$

$(49 \div 7) + (20 \div 4) - (42 \div 7) = \ldots$ $(28 \div 7) \times (8 \div 4) = \ldots$

Triumph points - …   Rule-breaking points - …

1. <u>Find</u> and <u>circle</u> or <u>cross out</u> the words to find out more about Severus Snape.

```
K J G Z Z D O A P I Q O R D P E I
D R Q U A G U A S N P G E F O C L
I P O T I O N S M A S T E R T N H
R N P W O D S P N W A R N J I E M
N G S O R S A Z X C B O E Q O I L
S K B U L E J E I P R F F M N C L
X L F X F Y P T H D Y F J D M S A
K G M K T F S A L R X R R P A E T
M D W S F I E U P Z E E H O K L I
O L U V H P A R Y S A D F H I T W
S D T P N C O C A D U P N B N B O
F P O S V R K H I B O O K U G U N
K S V R D D Z N G L L N I L D S K
K S S Y W Q G J V U N E E D N V D
H O G N I V A W D N A W O R E R Q
Y C I N V A D E R Z A C R D G T R
P A Y T C M Z U T K W H W I L W L
```

MIND-READING   INVADER

INSUFFERABLE   KNOW-IT-ALL

SUBTLE SCIENCE   POTION-MAKING

WAND-WAVING   CAULDRON

DUNDERHEAD   SOPHISTICATED

TEDIOUS PAPERWORK   POTIONS MASTER

Triumph points - …   Rule-breaking points - …

1. <u>Multiply</u>.

$$\begin{array}{r}90\\\times5\\\hline\end{array}\qquad\begin{array}{r}40\\\times8\\\hline\end{array}\qquad\begin{array}{r}50\\\times2\\\hline\end{array}\qquad\begin{array}{r}40\\\times4\\\hline\end{array}\qquad\begin{array}{r}70\\\times9\\\hline\end{array}$$

$$\begin{array}{r}50\\\times4\\\hline\end{array}\qquad\begin{array}{r}60\\\times7\\\hline\end{array}\qquad\begin{array}{r}40\\\times3\\\hline\end{array}\qquad\begin{array}{r}20\\\times7\\\hline\end{array}\qquad\begin{array}{r}80\\\times6\\\hline\end{array}$$

$$\begin{array}{r}30\\\times5\\\hline\end{array}\qquad\begin{array}{r}60\\\times8\\\hline\end{array}\qquad\begin{array}{r}50\\\times5\\\hline\end{array}\qquad\begin{array}{r}40\\\times6\\\hline\end{array}\qquad\begin{array}{r}30\\\times8\\\hline\end{array}$$

2. <u>Answer</u> the questions.

How many multiples of 7 are between 6 and 66? _____

How many multiples of 4 are between 2 and 33? _____

How many multiples of 9 are between 10 and 85? _____

How many multiples of 6 are between 4 and 59? _____

How many multiples of 3 are between 1 and 31? _____

How many multiples of 8 are between 5 and 52? _____

Triumph points - … Rule-breaking points - …

1. <u>Divide</u>.

$2\overline{)100}$ $3\overline{)150}$ $4\overline{)280}$ $5\overline{)100}$

$8\overline{)640}$ $5\overline{)300}$ $7\overline{)210}$ $6\overline{)420}$

$7\overline{)630}$ $8\overline{)720}$ $9\overline{)540}$ $8\overline{)160}$

2. Four cages contained 48, 67, 59, and 26 pixies respectively. <u>How could Professor Lockhart rearrange</u> the pixies so that each cage contained the same number of pixies?

Answer: _____.

Triumph points - …   Rule-breaking points - …

1. <u>Write in</u> the missing numbers on a potion factor tree. The number shows how many bottles you need, and the bottle shows how many grams of herbs you need to make a potion.

Football-Winning Potion

62 = __ × __

Football-Losing Potion

90 = __ × __ × __ × __

Behavior Potion

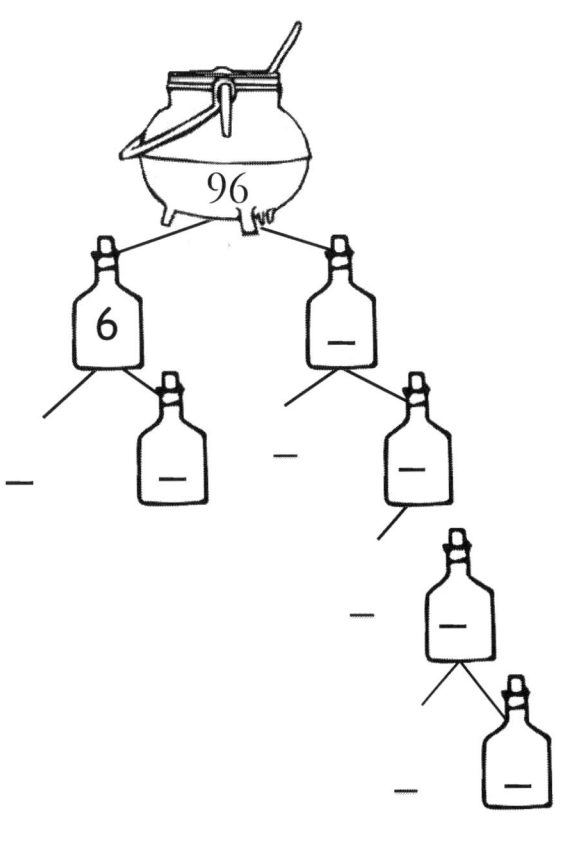

96 = __ × __ × __ × __ × __ × __

Peace-Disturbing Potion

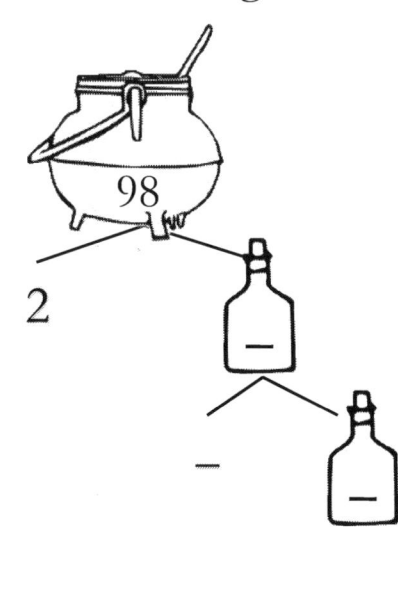

98 = __ × __ × __

Triumph points - …

Rule-breaking points - …

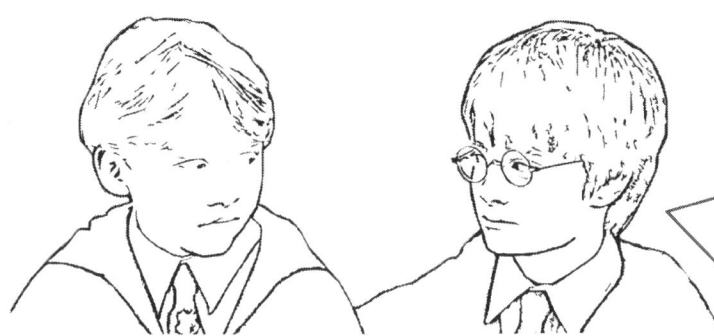

1. Answer the questions. Write one addition and multiplication number sentence for each problem. Indicate the order of operations.

Say you're ill... Really break your leg (see ***Harry Potter and the Sorcerer's Stone*** page 176).

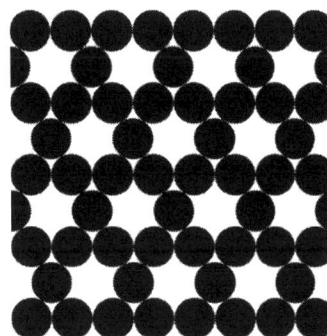

How many black circles are there?

____

____

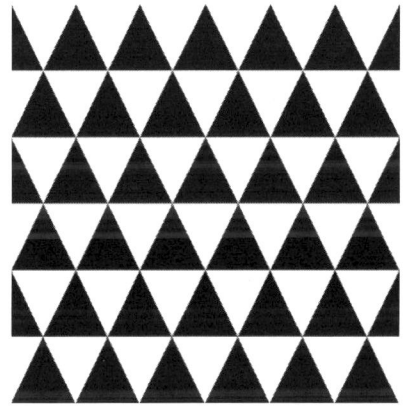

How many black triangles are there?

____

____

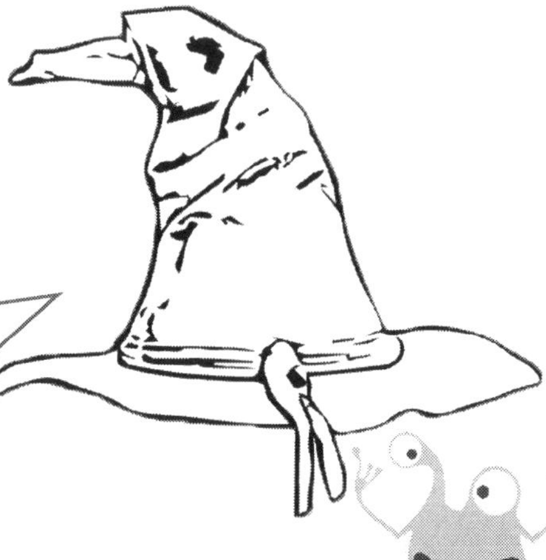

2. Find the 😢.

Just do your best,

(see ***Harry Potter and the Sorcerer's Stone*** page 105).

(87 + 63) ÷ 😢 = 30

____

😢 = ________________

(80 + 😢) ÷ 4 = 30

____

😢 = ________________

Triumph points - ...

Rule-breaking points - ...

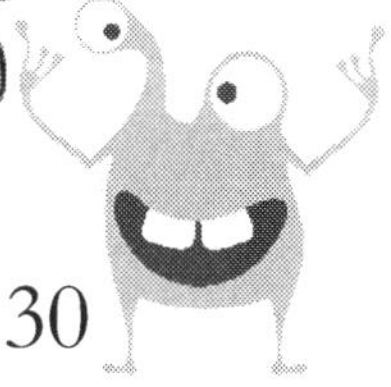

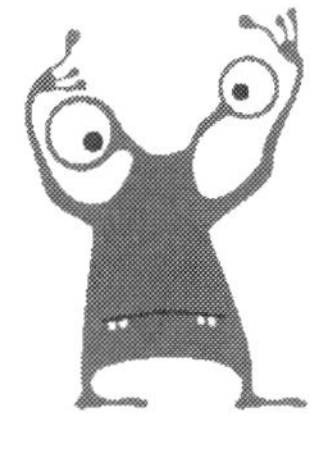

1. <u>Answer</u> the questions.

1. How many 30's are in:

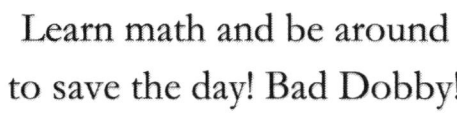
Learn math and be around to save the day! Bad Dobby!

240: 240 ÷ 30 = 8   180:   … ÷ … = …

360: … ÷ … = …   120:   … ÷ … = …

270: … ÷ … = …   210:   … ÷ … = …

How many 40's are in the highlighted numbers?

240: 240 ÷ 40 = …   320:   … ÷ … = …

360: … ÷ … = …   280:   … ÷ … = …

200: … ÷ … = …   160:   … ÷ … = …

2. <u>Multiply</u> and <u>find</u> the product. The first factor is ⬚six⬚ bronze Knuts.

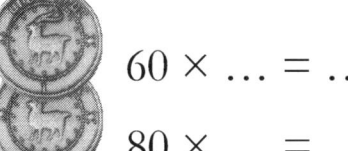

90 × 6 = 540   60 × … = …

50 × … = …   80 × … = …

40 × … = …

20 × … = …   30 × … = …

70 × … = …

3. <u>Divide</u> and <u>find</u> the quotient. The divisor is ⬚six⬚ bronze Knuts.

180 ÷ 6 = 30   480 ÷ … = …

420 ÷ … = …   300 ÷ … = …

240 ÷ … = …   540 ÷ … = …

360 ÷ … = …   120 ÷ … = …

Triumph points - …   Rule-breaking points - …

1. <u>Find</u> and <u>circle</u> or <u>cross out</u> the words to find out more about Ron Weasley.

```
E X H R D P V I M X B M V Q P P
R U D E A N T H O M A S J W G I
J O M Y U Q H M N P C G Y M O V
R E P E E K H C T I D D I U Q R
G Q W Z H U L Y K R R M F T X A
P J V C U A M P Z E S M V Y U D
S R E B B A C S S Z O T Z F R T
O G G R I M M A U L D P L A C H
D C A N L B H T L X Z R M J R U
W L D M M C Y Y H O W L E R O R
T M F X R Q W E V Z B G R F H W
Q Z I S S E H C D R A Z I W Z E
W I N G A R D I U M L A V I O A
G Y M S Y R R D J J A B H A B S
G U L Y B F S D D O K D V K G L
R E I L K L A S A V L W K I A E
Y V J A U F U R R I G Y A U J Y
```

MOLLY WEASLEY   ARTHUR WEASLEY

CHASER   DEAN THOMAS

GRIMMAULD PLACE   HORCRUX

WIZARD CHESS   HOWLER

QUIDDITCH KEEPER   SCABBERS

WINGARDIUM LAVIOSA   PIGWIDGEON

Triumph points - …   Rule-breaking points - …

Caution! My nightmare is on the rampage!

Oy, pea-brain!

(see **Harry Potter and the Sorcerer's Stone** page 143)

Oh, no (see **Harry Potter and the Sorcerer's Stone** page 105)

but Math first, so, <u>P</u>lace the parentheses to make the expressions true.

5 × 9 ÷ 3 = 15

50 ÷ 2 × 3 = 75

18 + 6 ÷ 3 = 8

3 × 10 ÷ 15 ÷ 3 = 6

Triumph points - …

Rule-breaking points - …

1. <u>Ansssswer</u> the questionsss.

1. <u>How many 50's</u> are in the highlighted numbers?

350: 350 ÷ 50 = …   300:   … ÷ … = …

450: … ÷ … = …   250:   … ÷ … = …

200: … ÷ … = …   400:   … ÷ … = …

<u>How many 60's</u> are in the highlighted numbers?

600: … ÷ … = …   300:   … ÷ … = …

480: … ÷ … = …   240:   … ÷ … = …

540: … ÷ … = …   420:   … ÷ … = …

2. <u>Multiply</u> and <u>find</u> the product. The first factor is seven bronze Knuts.

90 × 7 = 630   50 × … = …

40 × … = …   60 × … = …

20 × … = …   70 × … = …   30 × … = …   80 × … = …

3. <u>Divide</u> and <u>find</u> the quotient. The divisor is seven bronze Knuts.

490 ÷ … = …   140 ÷ 7 = 20

420 ÷ … = …   350 ÷ … = …

210 ÷ … = …

280 ÷ … = …   630 ÷ … = …   560 ÷ … = …

Triumph points - …   Rule-breaking points - …

The Unofficial Harry Potter Coloring Math Book Multiplication & Division Ages 8+

1. Circle the right answer.

A	90 × 2 × 2	A is equal to B
B	6 × 45	A is greater than B
		A is less than B

A	(20 × 40) - 250	A is greater than B
B	(40 × 20) + 250	A is equal to B
		A is less than B

A	(80 × 7) + 325	A is greater than B
B	(20 × 40) + 85	A is equal to B
		A is less than B

A	4 × 8 × 2 × 5	A is greater than B
B	4 × 4 × 4 × 6	A is less than B
		A is equal to B

Triumph points - …

Rule-breaking points - …

© 2019 STEM mindset, LLC   www.math-stepbystep.com

1. <u>Answer</u> the questions.

How many 70's are in the highlighted numbers?

700: ... ÷ ... = ...   350:   ... ÷ ... = ...

490: ... ÷ ... = ...   280:   ... ÷ ... = ...

420: ... ÷ ... = ...   210:   ... ÷ ... = ...

How many 80's are in the highlighted numbers?

800: ... ÷ ... = ...   400:   ... ÷ ... = ...

720: ... ÷ ... = ...   560:   ... ÷ ... = ...

640: ... ÷ ... = ...   320:   ... ÷ ... = ...

2. <u>Multiply</u> and <u>find</u> the product. The first factor is eight bronze Knuts.

90 × 8 = 720

50 × ... = ...

40 × ... = ...   60 × ... = ...

20 × ... = ...   70 × ... = ...   30 × ... = ...   80 × ... = ...

3. <u>Divide</u> and <u>find</u> the quotient. The divisor is eight bronze Knuts.

160 ÷ 8 = 20   480 ÷ ... = ...

400 ÷ ... = ...

640 ÷ ... = ...   720 ÷ ... = ...

320 ÷ ... = ...   240 ÷ ... = ...   560 ÷ ... = ...

Triumph points - ...   Rule-breaking points - ...

1. <u>Multiply</u>. <u>Use</u> five multiplication ways for each problem (<u>decompose</u> the factors).

24 × 4 = _____

_____

30 × 6 = _____

_____

45 × 8 = _____

_____

50 × 4 = _____

_____

2. <u>Help</u> me get the Triwizard Cup. Broomstick, no! Down!

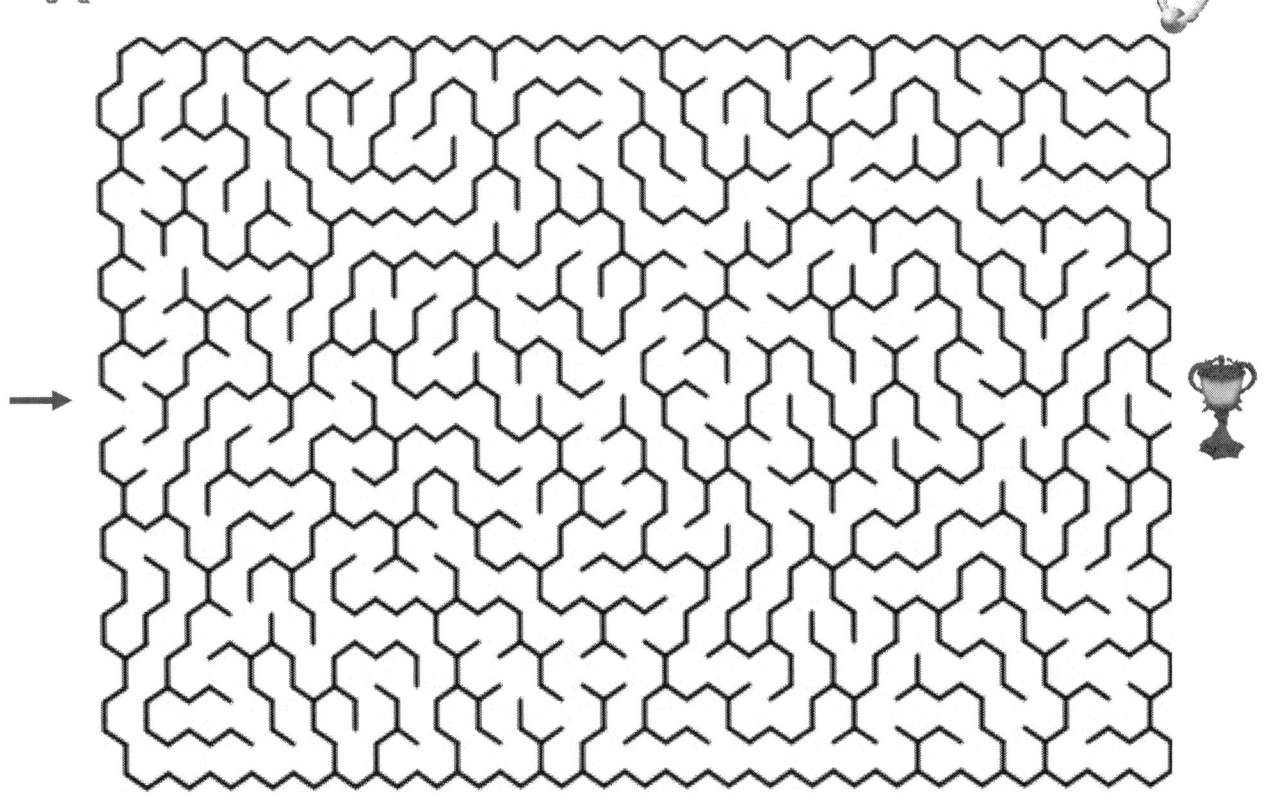

Triumph points - …

Rule-breaking points - …

1. How many 90's are in the highlighted numbers?

900: … ÷ … = …   450: … ÷ … = …

810: … ÷ … = …   540: … ÷ … = …

630: … ÷ … = …   720: … ÷ … = …

2. Multiply and find the product. The first factor is nine bronze Knuts.

60 × … = …

50 × … = …   90 × 9 = 810   80 × … = …

40 × … = …

20 × … = …   70 × … = …   30 × … = …

3. Divide and find the quotient. The divisor is nine bronze Knuts.

450 ÷ … = …   180 ÷ 9 = 20   720 ÷ … = …

540 ÷ … = …   630 ÷ … = …

270 ÷ … = …   360 ÷ … = …   810 ÷ … = …

4. Find the ☺.

(120 + 240) ÷ ☺ = 90   (119 + ☺) ÷ 5 = 70

_____   _____

☺ = _____   ☺ = _____

Triumph points - …   Rule-breaking points - …

1. <u>Find</u> and <u>circle</u> or <u>cross out</u> the words to find out more about Professor McGonagall.

```
G X L I G X K G Q U T M H A B U D H
U L A N Q N H U T E W Z N Y J O L T
I L C E K A U Y M I V I A D J L R Z
N E K P E P R B N F M C H M O A F W
E P O T T K K J T A R E Q R N Y F Z
A S F I M E T O G H G V T S G U L V
P G C T Y E G U V D G N F U S N L G
I N O U P T S V T N I I T X R Q R I
G I N D G B I V H A G F T P J N X S
S H F E T X Q Y T U Y T P Z E K E L
L C I S A A T N R L L A B E L U Y R
X T D O M G U A D X M T N G F K G H
P I E R T O T U M L O C O M O T O R
V W N P M I N E E D L E W O R K R D
Z S C H O P W P K P A Z V T P F E Q
G R E N L L E P S G N I H S I N A V
G O K H A V E A B I S C U I T C P R
```

VANISHING SPELL   TRANSFIGURATION

ANIMAGUS   NEEDLEWORK

TIME-TURNER   PIERTOTUM LOCOMOTOR

SWITCHING SPELL   GUINEA-PIGS

YULE BALL   TIGHT BUN

INEPTITUDES   HAVE A BISCUIT

Triumph points - …   Rule-breaking points - …

1. <u>Multiply</u>.

```
    40        50        60        90        90
×    5    ×    8    ×    2    ×    4    ×    9
───────   ───────   ───────   ───────   ───────

    60        50        50        60        30
×    4    ×    7    ×    3    ×    7    ×    6
───────   ───────   ───────   ───────   ───────

    80        30        90        40        80
×    5    ×    8    ×    5    ×    6    ×    8
───────   ───────   ───────   ───────   ───────
   400
```

2. Harry counted ⟦9⟧ kittens (k) and owls (o) outside. The kittens and birds had ⟦28⟧ legs altogether. <u>How many owls</u> did Harry see?

Kittens & Birds	Legs	Kittens	Owls
___ ___	___	? ___	? ___

Answer: _____

Triumph points - …   Rule-breaking points - …

1. <u>Divide</u>.

$$2\overline{)200} \quad 3\overline{)270} \quad 8\overline{)400} \quad 8\overline{)400}$$

$$8\overline{)480} \quad 5\overline{)450} \quad 7\overline{)490} \quad 6\overline{)300}$$

$$7\overline{)420} \quad 8\overline{)560} \quad 9\overline{)540} \quad 8\overline{)720}$$

2. Ron had 70 Chocolate Frog cards. After he gave Harry 7 cards, Ron had nine times as many cards as Harry. <u>How many cards</u> did Harry have?

Answer: _____.

Triumph points - …   Rule-breaking points - …

1. <u>Write in</u> the missing numbers on a potion factor tree. The number shows how many bottles you need, and the bottle shows how many grams of herbs you need to make a potion.

Extendable Ears Potion

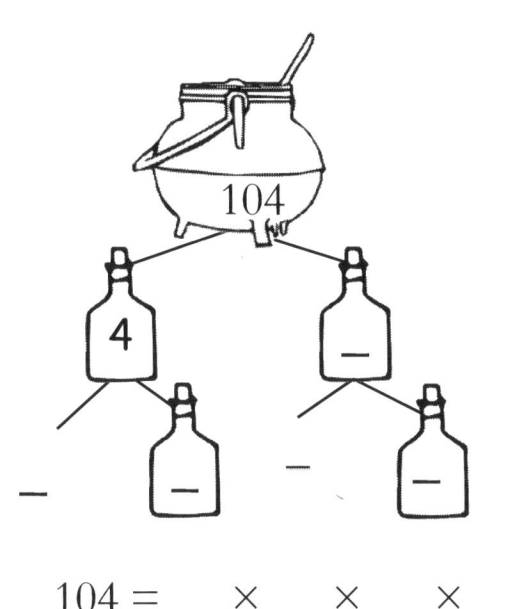

$104 = \_\_ \times \_\_ \times \_\_ \times \_\_$

Extendable Ears Potion Antidote

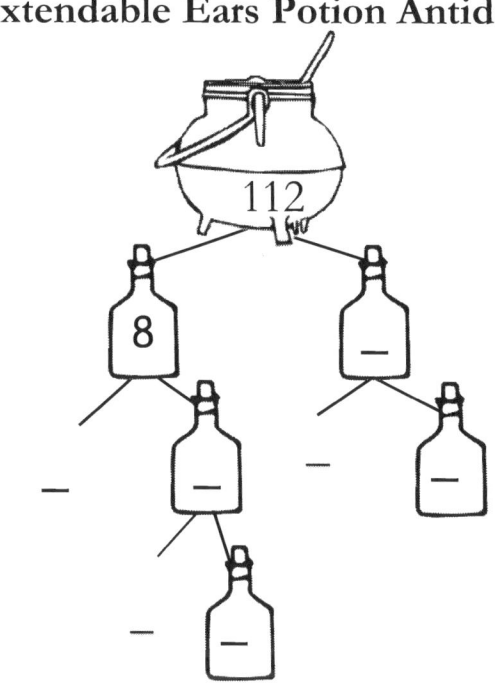

$112 = \_\_ \times \_\_ \times \_\_ \times \_\_ \times \_\_$

Bruise-Healing Paste

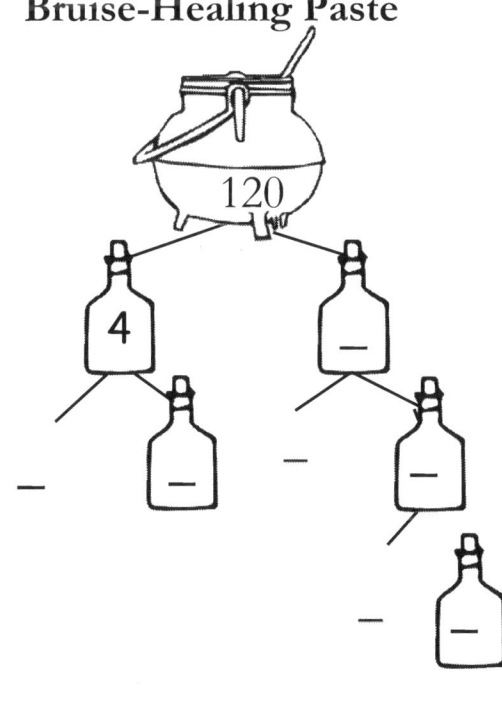

$102 = \_\_ \times \_\_ \times \_\_ \times \_\_ \times \_\_$

Essence of Wonder

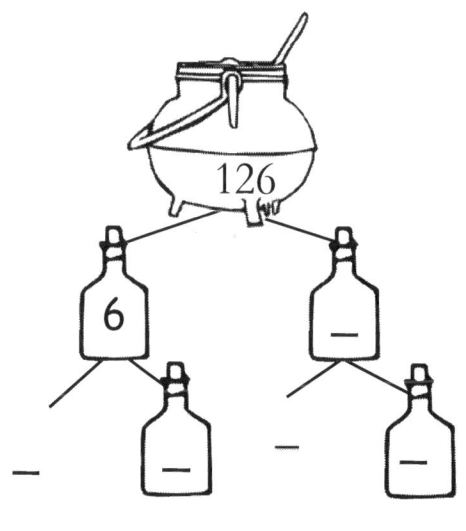

$126 = \_\_ \times \_\_ \times \_\_ \times \_\_$

Triumph points - …

Rule-breaking points - …

1. Place the parentheses to make the expressions true.

$7 \times 8 \div 4 = 14$ $30 \div 3 \times 2 = 5$

$10 + 4 \div 7 = 2$ $6 \times 2 \times 5 \div 2 = 30$

CHAPTER 1

Fun with

Multiplying and Dividing within 100 without Regrouping

Multiplying by 2-9s using arrays

Learning Multiplication Facts

Dividing by 2-9s

Learning the Order of Operations

Word Problems

Factors and Multiples

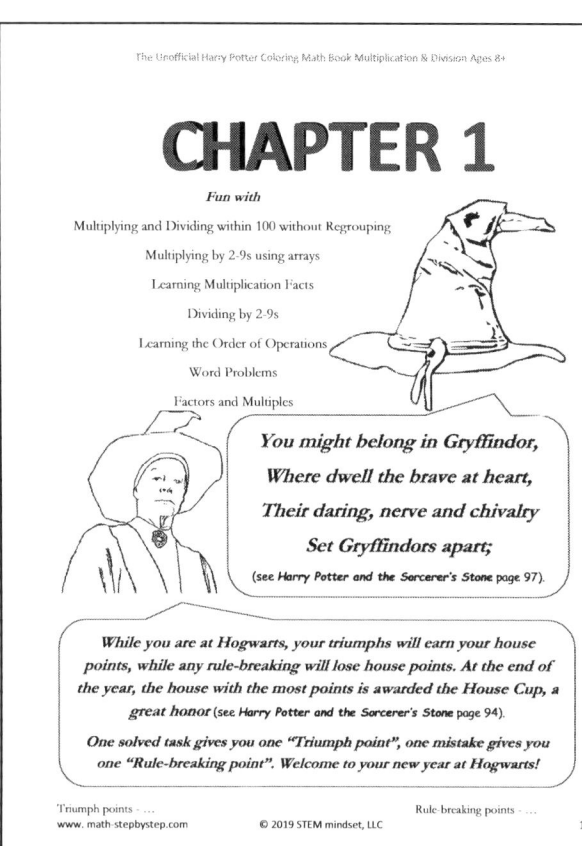

*You might belong in Gryffindor,
Where dwell the brave at heart,
Their daring, nerve and chivalry
Set Gryffindors apart;*
(see *Harry Potter and the Sorcerer's Stone* page 97).

While you are at Hogwarts, your triumphs will earn your house points, while any rule-breaking will lose house points. At the end of the year, the house with the most points is awarded the House Cup, a great honor (see *Harry Potter and the Sorcerer's Stone* page 94).

One solved task gives you one "Triumph point", one mistake gives you one "Rule-breaking point". Welcome to your new year at Hogwarts!

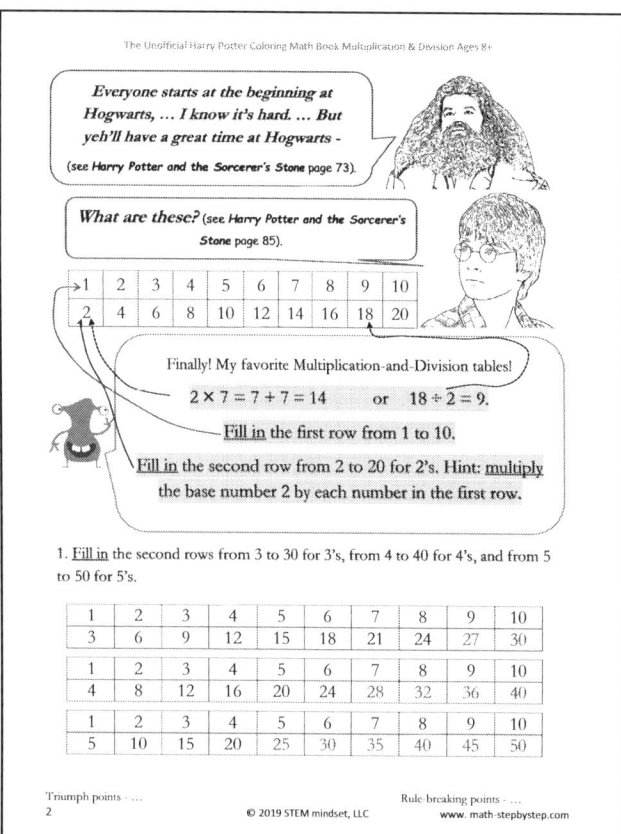

Everyone starts at the beginning at Hogwarts, ... I know it's hard. ... But yeh'll have a great time at Hogwarts - (see *Harry Potter and the Sorcerer's Stone* page 73).

What are these? (see *Harry Potter and the Sorcerer's Stone* page 85).

1	2	3	4	5	6	7	8	9	10
2	4	6	8	10	12	14	16	18	20

Finally! My favorite Multiplication-and-Division tables!

$2 \times 7 = 7 + 7 = 14$ or $18 \div 2 = 9$.

Fill in the first row from 1 to 10.

Fill in the second row from 2 to 20 for 2's. Hint: multiply the base number 2 by each number in the first row.

1. Fill in the second rows from 3 to 30 for 3's, from 4 to 40 for 4's, and from 5 to 50 for 5's.

1	2	3	4	5	6	7	8	9	10
3	6	9	12	15	18	21	24	27	30

1	2	3	4	5	6	7	8	9	10
4	8	12	16	20	24	28	32	36	40

1	2	3	4	5	6	7	8	9	10
5	10	15	20	25	30	35	40	45	50

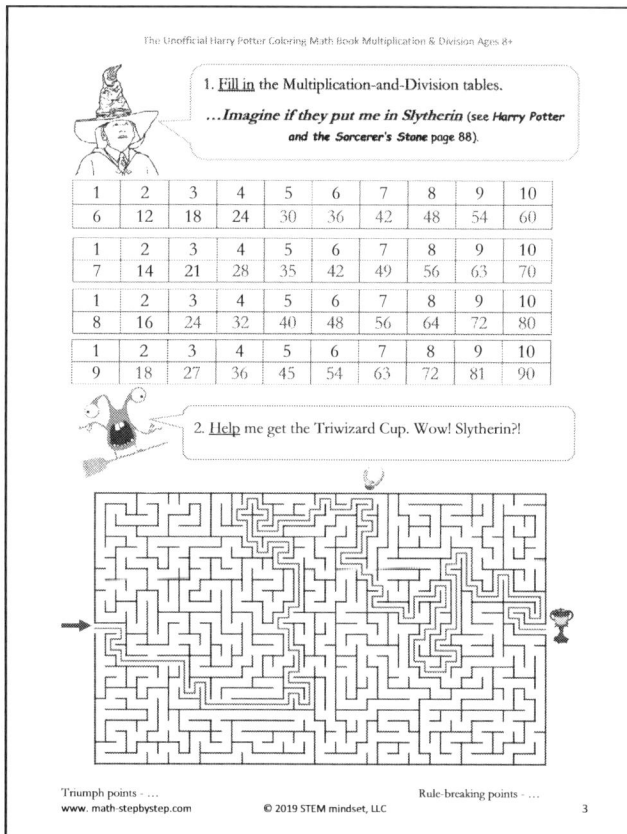

1. Fill in the Multiplication-and-Division tables.

...Imagine if they put me in Slytherin (see *Harry Potter and the Sorcerer's Stone* page 88).

1	2	3	4	5	6	7	8	9	10
6	12	18	24	30	36	42	48	54	60

1	2	3	4	5	6	7	8	9	10
7	14	21	28	35	42	49	56	63	70

1	2	3	4	5	6	7	8	9	10
8	16	24	32	40	48	56	64	72	80

1	2	3	4	5	6	7	8	9	10
9	18	27	36	45	54	63	72	81	90

2. Help me get the Triwizard Cup. Wow! Slytherin?!

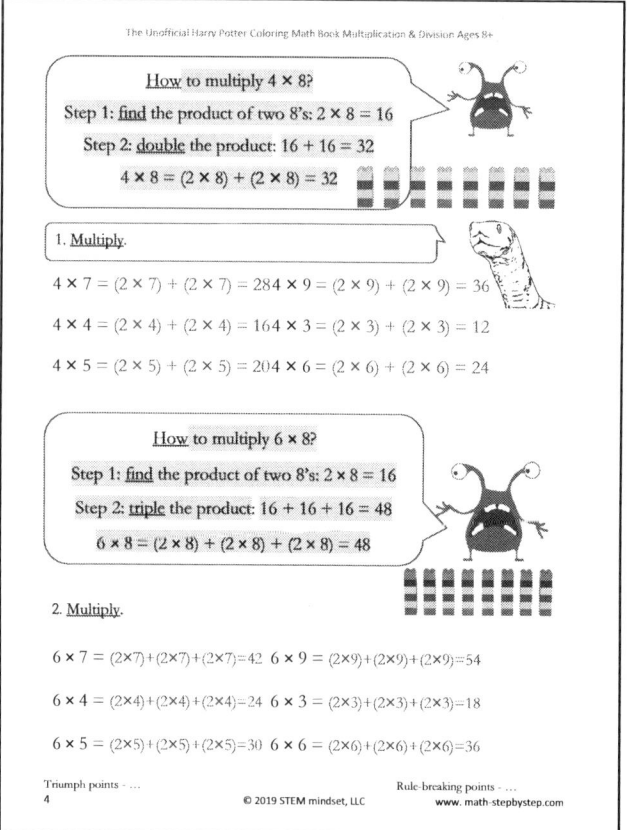

How to multiply 4×8?

Step 1: find the product of two 8's: $2 \times 8 = 16$

Step 2: double the product: $16 + 16 = 32$

$4 \times 8 = (2 \times 8) + (2 \times 8) = 32$

1. Multiply.

$4 \times 7 = (2 \times 7) + (2 \times 7) = 28$ $4 \times 9 = (2 \times 9) + (2 \times 9) = 36$

$4 \times 4 = (2 \times 4) + (2 \times 4) = 16$ $4 \times 3 = (2 \times 3) + (2 \times 3) = 12$

$4 \times 5 = (2 \times 5) + (2 \times 5) = 20$ $4 \times 6 = (2 \times 6) + (2 \times 6) = 24$

How to multiply 6×8?

Step 1: find the product of two 8's: $2 \times 8 = 16$

Step 2: triple the product: $16 + 16 + 16 = 48$

$6 \times 8 = (2 \times 8) + (2 \times 8) + (2 \times 8) = 48$

2. Multiply.

$6 \times 7 = (2 \times 7) + (2 \times 7) + (2 \times 7) = 42$ $6 \times 9 = (2 \times 9) + (2 \times 9) + (2 \times 9) = 54$

$6 \times 4 = (2 \times 4) + (2 \times 4) + (2 \times 4) = 24$ $6 \times 3 = (2 \times 3) + (2 \times 3) + (2 \times 3) = 18$

$6 \times 5 = (2 \times 5) + (2 \times 5) + (2 \times 5) = 30$ $6 \times 6 = (2 \times 6) + (2 \times 6) + (2 \times 6) = 36$

1. <u>Write in</u> the missing numbers on a potion factor tree. The number shows how many bottles you need, and the bottle shows how many grams of herbs you need to make a potion.

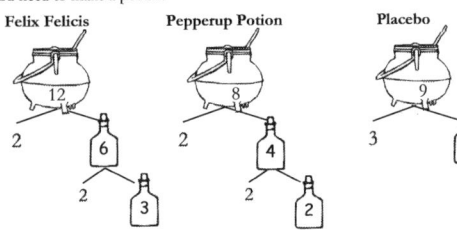

Felix Felicis **Pepperup Potion** **Placebo**

Factors are numbers that, when multiplied together (for example, 2 × 5) form a new number (for example, 10) called a product.

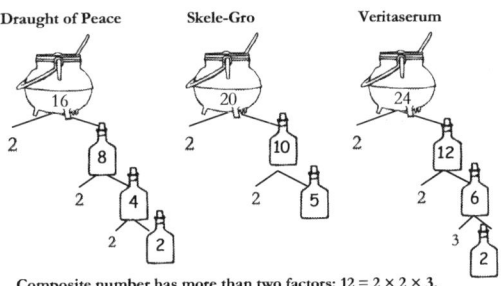

Draught of Peace **Skele-Gro** **Veritaserum**

Composite number has more than two factors: 12 = 2 × 2 × 3.

<u>How</u> to multiply 7 × 8? Hint: when the multiplier (the first factor) is 1 less than the multiplicand (the second factor) you need to square the multiplicand (multiply the number by itself) and then, subtract the multiplicand out of the product:

Step 1: <u>find</u> the product of eight 8's: 8 × 8 = 64

Step 2: <u>subtract</u> one 8 out of the product: 64 - 8 = 56

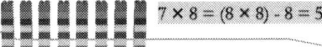 7 × 8 = (8 × 8) - 8 = 56

1. <u>Multiply</u>. The multiplier (the first factor) is 1 less than the multiplicand (the second factor).

6 × 7 = (7 × 7) − 7 = 42   8 × 9 = (9 × 9) − 9 = 72

3 × 4 = (4 × 4) − 4 = 12   2 × 3 = (3 × 3) − 3 = 6

4 × 5 = (5 × 5) − 5 = 20   9 × 10 = (10 × 10) − 10 = 90

2. There are 35 pixies in all. I divided the pixies into groups of 5. <u>How many small pixies</u> are there?

Answer: (35÷5) = 7 (groups)   (4×2)+(3×3)= 17 (small pixies).

How to multiply 6 × 5? Hint: when the multiplier (the first factor) is 1 more than the multiplicand (the second factor) you need to square the multiplicand (multiply the number by itself) and then, add one more multiplicand to the product:

Step 1: <u>find</u> the product of five 5's: 5 × 5 = 25

Step 2: <u>add</u> one 5 to the product: 25 + 5 = 30

6 × 5 = (5 × 5) + 5 = 30

1. <u>Multiply</u>. The multiplier (the first factor) is 1 more than the multiplicand (the second factor).

5 × 4 = (4 × 4) + 4 = 20   8 × 7 = (7 × 7) + 7 = 56

3 × 2 = (2 × 2) + 2 = 6   4 × 3 = (3 × 3) + 3 = 12

9 × 8 = (8 × 8) + 8 = 72   7 × 6 = (6 × 6) + 6 = 42

2. There are 43 pixies in all. I divided the pixies into groups of 4. <u>How many big pixies</u> are there?

Answer: (40 ÷ 4) × 2 + 1 = 21 (big pixies).

How to multiply 9 × 8?

Step 1: <u>find</u> the product of ten 8's: 10 × 8 = 80

Step 2: <u>subtract</u> one 8 out of the product: 80 - 8 = 72

9 × 8 = (10 × 8) - 8 = 72

1. <u>Multiply</u>.

9 × 7 = (10 × 7) - 7 = 63   9 × 9 = (10 × 9) - 9 = 81

9 × 4 = (10 × 4) - 4 = 36   9 × 3 = (10 × 3) - 3 = 27

9 × 5 = (10 × 5) - 5 = 45   9 × 6 = (10 × 6) - 6 = 54

How to multiply 5 × 8?

Hint: the best way to multiply by 5 is to multiply a number by 10 and divide the product by 2 since 5 = 10 ÷ 2

Step 1: <u>find</u> the product of ten 8's: 10 × 8 = 80

Step 2: <u>divide</u> the product by 2: 80 ÷ 2 = 40

5 × 8 = (10 × 8) ÷ 2 = 40

2. <u>Multiply</u>.

5 × 7 = (10 × 7) ÷ 2 = 35   5 × 9 = (10 × 9) ÷ 2 = 45

5 × 4 = (10 × 4) ÷ 2 = 20   5 × 3 = (10 × 3) ÷ 2 = 15

5 × 5 = (10 × 5) ÷ 2 = 25   5 × 6 = (10 × 6) ÷ 2 = 30

The Unofficial Harry Potter Coloring Math Book Multiplication & Division Ages 8+

1. <u>Find</u> and circle or cross out the words to find out more about Harry Potter.

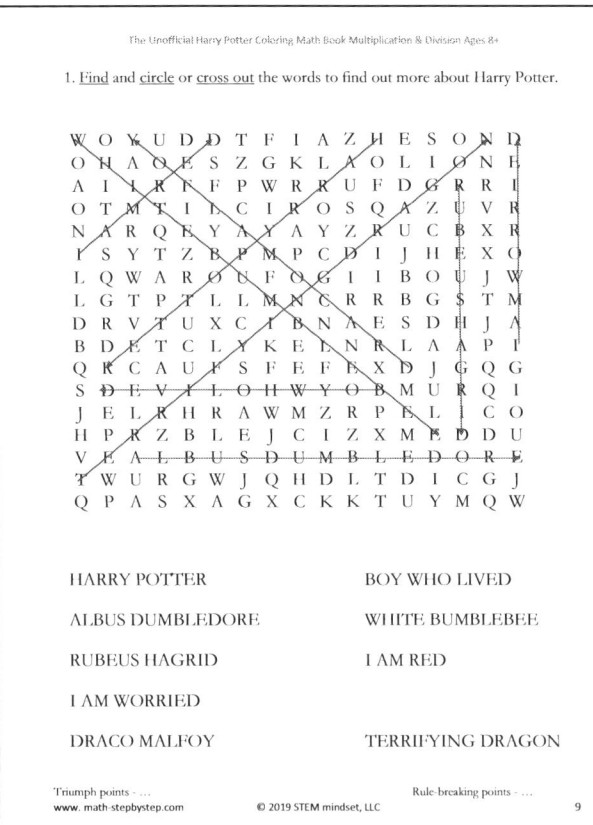

HARRY POTTER

ALBUS DUMBLEDORE

RUBEUS HAGRID

I AM WORRIED

DRACO MALFOY

BOY WHO LIVED

WHITE BUMBLEBEE

I AM RED

TERRIFYING DRAGON

---

1. Look at the pattern below and answer the questionsss.

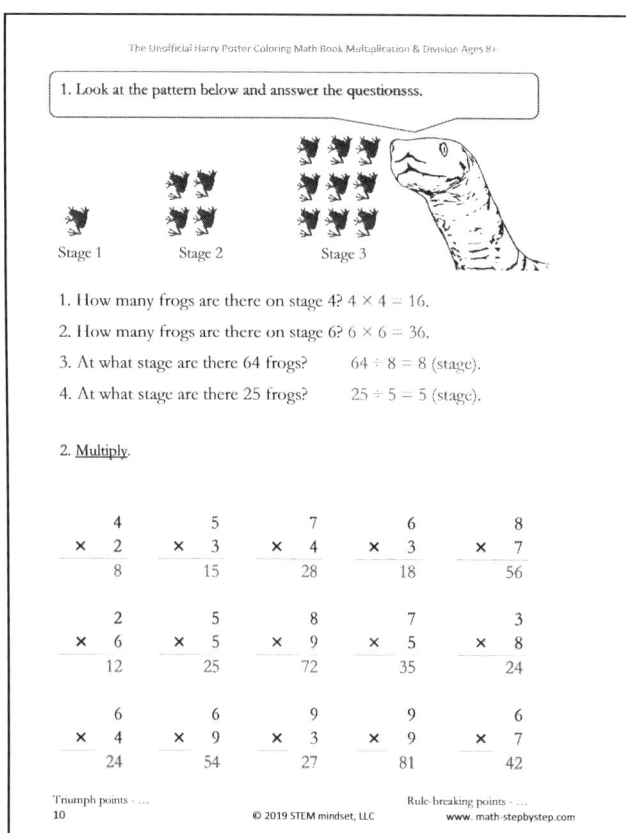

Stage 1   Stage 2   Stage 3

1. How many frogs are there on stage 4? $4 \times 4 = 16$.
2. How many frogs are there on stage 6? $6 \times 6 = 36$.
3. At what stage are there 64 frogs? $64 \div 8 = 8$ (stage).
4. At what stage are there 25 frogs? $25 \div 5 = 5$ (stage).

2. <u>Multiply</u>.

$4 \times 2 = 8$	$5 \times 3 = 15$	$7 \times 4 = 28$	$6 \times 3 = 18$	$8 \times 7 = 56$
$2 \times 6 = 12$	$5 \times 5 = 25$	$8 \times 9 = 72$	$7 \times 5 = 35$	$3 \times 8 = 24$
$6 \times 4 = 24$	$6 \times 9 = 54$	$9 \times 3 = 27$	$9 \times 9 = 81$	$6 \times 7 = 42$

---

1. Professor Lockhart showed two cages with pixies. The big cage had three times as many pixies as the small cage. If there were 27 pixies in the big cage, how many pixies were kept in the small cage?

Small cage	Big cage	Small cage
Unknown?	3×small 27	? 9

Big cage = 3 × small cage
27 = 3 × small cage
Small cage = 27 ÷ 3   Small cage = 9
Answer: 9 pixies were kept in the small cage.

2. <u>Write in</u> the missing numbers on a potion factor tree. The number shows how many bottles you need, and the bottle shows how many grams of herbs you need to make a potion.

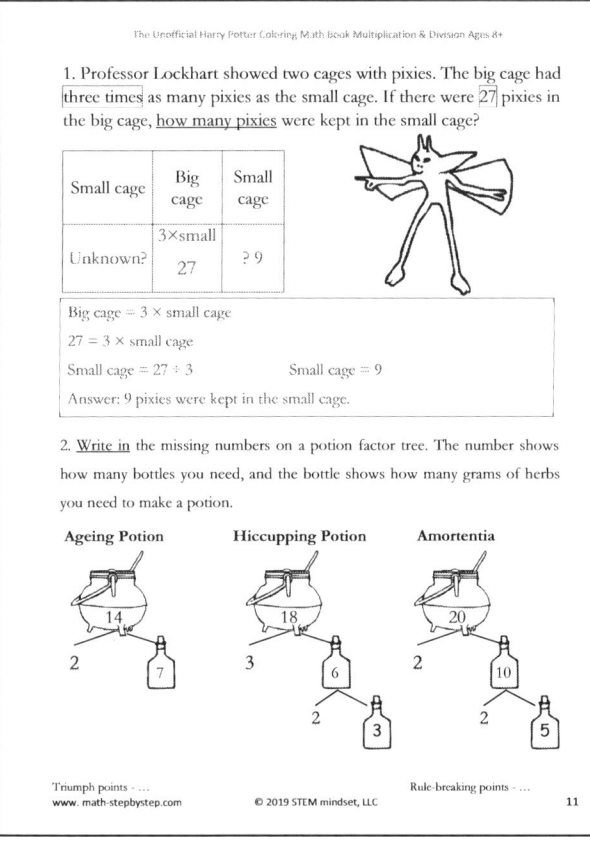

Ageing Potion: 14, 2, 7
Hiccupping Potion: 18, 3, 6, 2, 3
Amortentia: 20, 2, 10, 2, 5

Brainers, I will do division. I like tricks, may I start with the tricks?

We divide 6 by 2:
$6 \div 2 = 3$.

$6 \div 2 = \ldots$

6 divided by 2

Or I can rewrite the problem as: → $2\overline{)6}$

Step 1: divide 6 by 2: $6 \div 2 = 3$.
The answer is 3. → $2\overline{)6}$

Step 2: write 3 in one's place above the division sign. → $\begin{array}{r}3\\2\overline{)6}\end{array}$

Step 3: multiply to check your answer: $3 \times 2 = 6$. Write the product 6 under the dividend 6. → $\begin{array}{r}3\\2\overline{)6}\\-6\end{array}$

Step 4: subtract the product 6 from the dividend 6: $6 - 6 = 0$ → $\begin{array}{r}3\\2\overline{)6}\\-6\\\hline 0\end{array}$

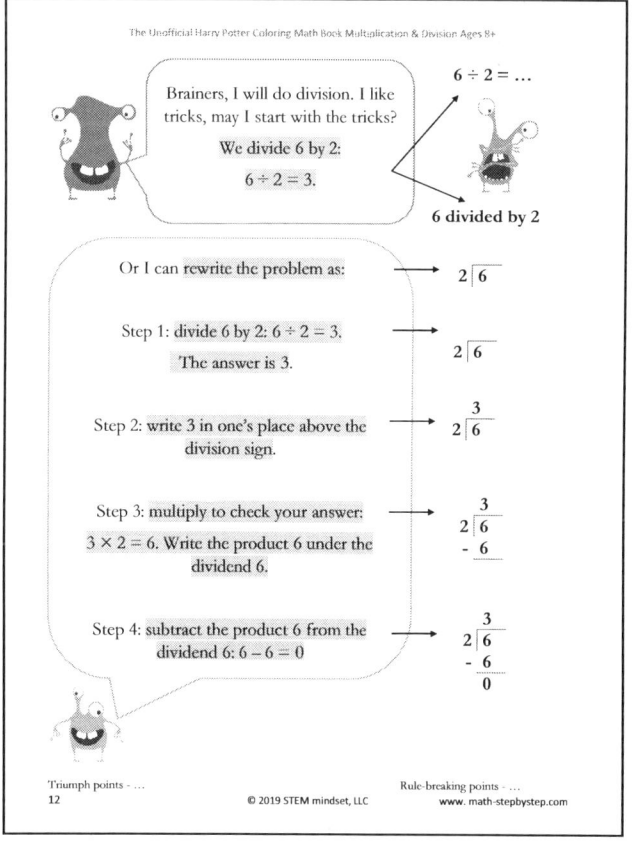

Page 13

1. <u>Divide</u>.

$$2\overline{)4} \quad 3\overline{)9} \quad 4\overline{)12} \quad 5\overline{)10} \quad 6\overline{)12}$$

$$8\overline{)16} \quad 5\overline{)15} \quad 7\overline{)14} \quad 6\overline{)18} \quad 5\overline{)25}$$

$$7\overline{)28} \quad 8\overline{)24} \quad 9\overline{)27} \quad 8\overline{)32} \quad 9\overline{)18}$$

2. Ron got <u>8</u> points. Hermione got <u>seven times</u> as many points as Ron. <u>How many points</u> did they get altogether?

Ron	Hermione	Hermione	Altogether
8	7 × Ron	? 56	? 64

Hermione: 7 × Ron = 7 × 8 = 56

Altogether: 8 + 56 = 64

Answer: 64 points.

Page 14

1. <u>Find</u> the value.

$2 \times 10 = 20$
$20 - 2 = 18$
$2 \times 9 = 18$

$5 \times 10 = 50$
$50 - 5 = 45$
$5 \times 9 = 45$

$8 \times 10 = 80$
$80 - 8 = 72$
$8 \times 9 = 72$

$3 \times 10 = 30$
$30 - 3 = 27$
$3 \times 9 = 27$

$9 \times 10 = 90$
$90 - 9 = 81$
$9 \times 9 = 81$

Page 15

1. <u>Multiply</u>. <u>Add</u> the products in each column. The Sorting Hat will choose the greatest sum. <u>Circle</u> the column with the greatest sum.

There's nothing hidden in your head The Sorting Hat can't see (see Harry Potter and the Sorcerer's Stone page 97).

Not Slytherin, not Slytherin (see Harry Potter and the Sorcerer's Stone page 100).

Slytherin	Hufflepuff	Gryffindor	Ravenclaw
7 × 4 = 28	8 × 5 = 40	6 × 3 = 18	5 × 5 = 25
3 × 4 = 12	7 × 2 = 14	9 × 3 = 27	2 × 8 = 16
3 × 7 = 21	4 × 6 = 24	5 × 3 = 15	8 × 7 = 56
9 × 5 = 45	8 × 4 = 32	9 × 9 = 81	8 × 9 = 72
9 × 6 = 54	8 × 6 = 48	7 × 7 = 49	4 × 2 = 8
5 × 4 = 20	6 × 5 = 30	2 × 5 = 10	8 × 5 = 40
2 × 6 = 12	7 × 5 = 35	3 × 6 = 18	6 × 6 = 36
2 × 9 = 18	4 × 9 = 36	8 × 3 = 24	7 × 3 = 21
4 × 4 = 16	6 × 4 = 24	5 × 5 = 25	5 × 3 = 15
7 × 6 = 42	9 × 7 = 63	2 × 2 = 4	8 × 8 = 64
268	346	271	353

Page 16

1. <u>Write in</u> the missing numbers on a potion factor tree. The number shows how many bottles you need, and the bottle shows how many grams of herbs you need to make a potion.

Antidote to Veritaserum: 28 → 2, 14 → 2, 7

Babbling Potion: 28 → 7, 4 → 2, 2

Calming Potion: 30 → 2, 15 → 3, 5

A prime number: is greater than 1; has factors only 1 and itself
2, 3, 5, 7, 11, 13, 17, 19, 23, 29, etc.

Polyjuice Potion: 36 → 2, 18 → 2, 9 → 3, 3

Dizziness Potion: 30 → 5, 6 → 2, 3

Confusing Concoction: 32 → 2, 16 → 2, 8 → 2, 4 → 2, 2

Page 17

1. <u>Find</u> the value.

I bet I'm the worst in the class (see *Harry Potter and the Sorcerer's Stone* page 83).

$6 \times 10 = 60$
$60 - 6 = 54$
$6 \times 9 = 54$

$4 \times 10 = 40$
$40 - 4 = 36$
$4 \times 9 = 36$

$7 \times 10 = 70$
$70 - 7 = 63$
$7 \times 9 = 63$

2. Evaluate each expression. <u>First, start with the parentheses (brackets), then, addition or subtraction from left to right!</u> <u>Indicate</u> the order of operation.

$\overset{1}{(7 \times 10)} + \overset{4}{} \overset{2}{(4 \times 9)} + \overset{5}{} \overset{3}{(8 \times 10)} = 186$ $\overset{1}{(7 \times 9)} + \overset{4}{} \overset{2}{(3 \times 10)} + \overset{5}{} \overset{3}{(8 \times 9)} = 165$

$(9 \times 9) + (3 \times 9) + (6 \times 10) = 168$ $(5 \times 9) - (2 \times 10) + (6 \times 9) = 79$

$(5 \times 10) - (4 \times 10) + (2 \times 9) = 28$ $(9 \times 10) + (5 \times 9) - (6 \times 10) = 75$

Page 18

1. <u>Answer</u> the questions.

You won't be. There's loads of people who … learn quick enough (see *Harry Potter and the Sorcerer's Stone* page 83).

I have a picture with black & white triangles. <u>How many white triangles</u> are there? Easy!

I use multiplication:
6 (down) × 3 (across) = 18 (white triangles).

Or I can add: $3 + 3 + 3 + 3 + 3 + 3 = 18$.

<u>How many white squares</u> are there? <u>Write</u> one addition and one multiplication number sentence.

$3 \times 3 = 9$

$3 + 3 + 3 = 9$

<u>How many white 4-pointed stars</u> are there?

$4 \times 4 = 16$

$4 + 4 + 4 + 4 = 16$

<u>How many black hexagons</u> are there?

$(4 \times 3) + (3 \times 4) = 12 + 12 = 24$

$3 + 3 + 3 + 3 + 4 + 4 + 4 = 24$

Page 19

Harry … could see the little round ball, wings fluttering, darting up ahead - he put on an extra spurt of speed - (see *Harry Potter and the Sorcerer's Stone* page 155).

1. <u>Answer</u> the questions.

Find a pair of one-digit numbers when the sum is 14 and the product is 45: $9 + 5 = 14, 9 \times 5 = 45$.

Find a pair of one-digit numbers when the sum is 16 and the product is 64: $8 + 8 = 16, 8 \times 8 = 64$.

Page 20

1. <u>Multiply</u> and <u>find</u> the value.

Brazil, here I come… Thanksss, amigo (see *Harry Potter and the Sorcerer's Stone* page 23).

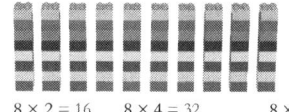

$8 \times 2 = 16$ $8 \times 4 = 32$ $8 \times 6 = 48$
$8 \times 8 = 64$ $8 \times 10 = 80$

$6 \times 2 = 12$ $6 \times 4 = 24$ $6 \times 6 = 36$
$6 \times 8 = 48$ $6 \times 10 = 60$

2. <u>Evaluate</u> each expression. <u>First, start with the parentheses (brackets), then, addition or subtraction from left to right!</u> <u>Indicate</u> the order of operation.

$\overset{1}{(8 \times 4)} + \overset{4}{} \overset{2}{(2 \times 2)} + \overset{5}{} \overset{3}{(5 \times 10)} = 86$ $\overset{1}{(2 \times 8)} + \overset{4}{} \overset{2}{(3 \times 6)} + \overset{5}{} \overset{3}{(8 \times 8)} = 98$

$(3 \times 10) + (2 \times 6) + (8 \times 2) = 58$ $(5 \times 6) - (2 \times 10) + (3 \times 4) = 22$

$(5 \times 8) - (2 \times 4) + (3 \times 8) = 56$ $(8 \times 10) + (5 \times 2) - (3 \times 2) = 84$

$(5 \times 4) - (2 \times 2) + (8 \times 6) = 64$ $(8 \times 4) + (5 \times 4) - (5 \times 10) = 2$

Page 21

1. <u>Write</u> in the missing numbers on a potion factor tree. The number shows how many bottles you need, and the bottle shows how many grams of herbs you need to make a potion.

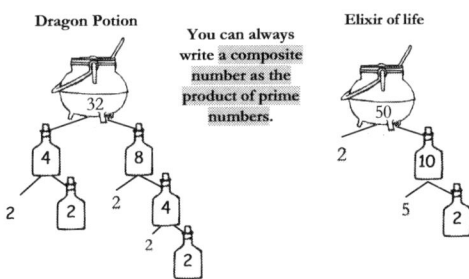

You can always write a composite number as the product of prime numbers.

2. To brew a Draught of Living Death Potion you need to add powdered root of asphodel to an infusion of wormhood. The powdered root of asphodel weighs <u>three times</u> as much as an infusion of wormhood. If the powdered root of asphodel weighs <u>24</u> grams, <u>how much</u> do they weigh altogether?

Asphodel	Wormhood	Altogether
3×W; 24	? 8	? 32

Asphodel = 3 × Wormhood
24 = 3 × Wormhood
Wormhood = 24 ÷ 3 = 8
Altogether: 24 + 8 = 32
Answer: 32 grams.

Page 22

1. <u>Answer</u> the questions. <u>Write</u> one multiplication and addition number sentence for each problem.

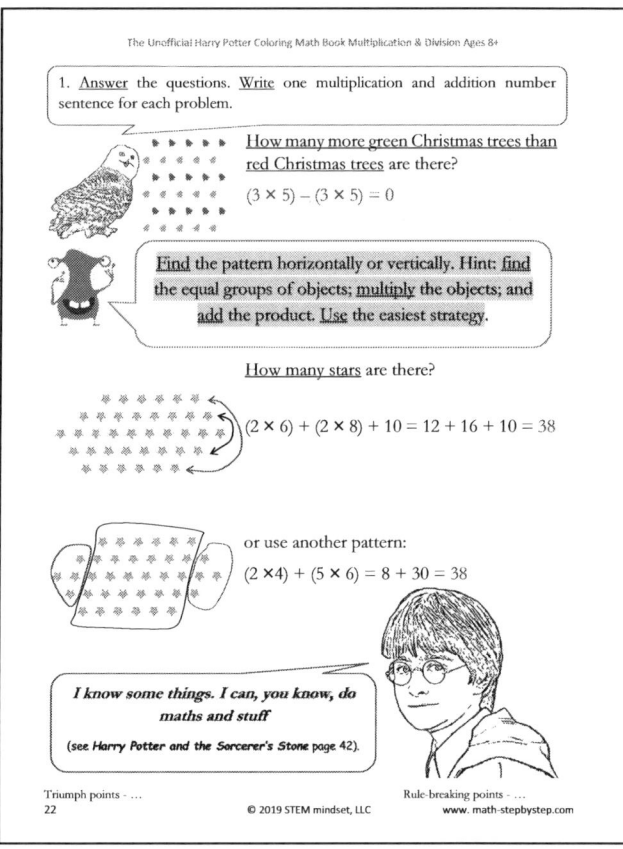

How many more green Christmas trees than red Christmas trees are there?
$(3 \times 5) - (3 \times 5) = 0$

<u>Find</u> the pattern horizontally or vertically. Hint: <u>find</u> the equal groups of objects; <u>multiply</u> the objects; and <u>add</u> the product. <u>Use</u> the easiest strategy.

How many stars are there?
$(2 \times 6) + (2 \times 8) + 10 = 12 + 16 + 10 = 38$

or use another pattern:
$(2 \times 4) + (5 \times 6) = 8 + 30 = 38$

I know some things. I can, you know, do maths and stuff
(see *Harry Potter and the Sorcerer's Stone* page 42).

Page 23

1. <u>Find</u> and <u>circle</u> or <u>cross out</u> the words to find out more about Harry Potter.

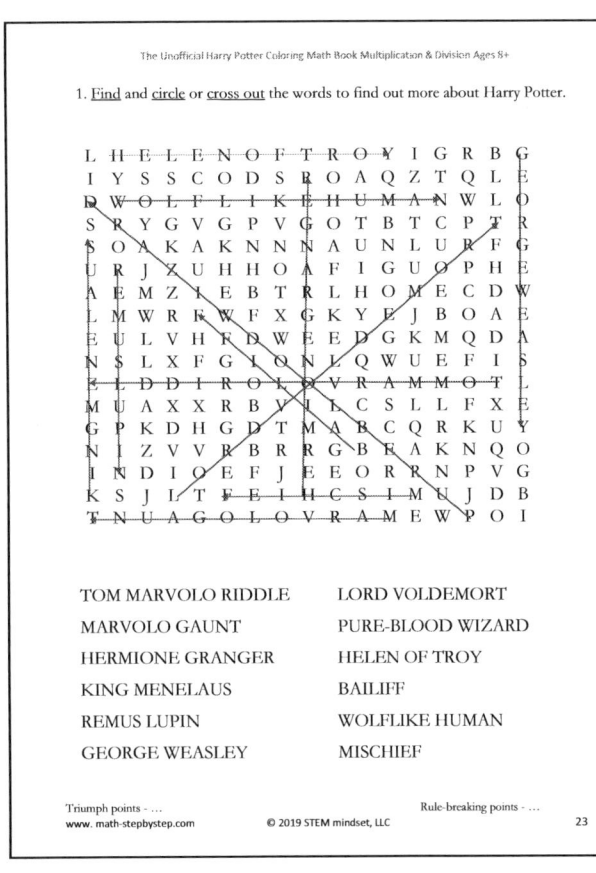

TOM MARVOLO RIDDLE — LORD VOLDEMORT
MARVOLO GAUNT — PURE-BLOOD WIZARD
HERMIONE GRANGER — HELEN OF TROY
KING MENELAUS — BAILIFF
REMUS LUPIN — WOLFLIKE HUMAN
GEORGE WEASLEY — MISCHIEF

Page 24

1. Look at the pattern below and answer the questions.

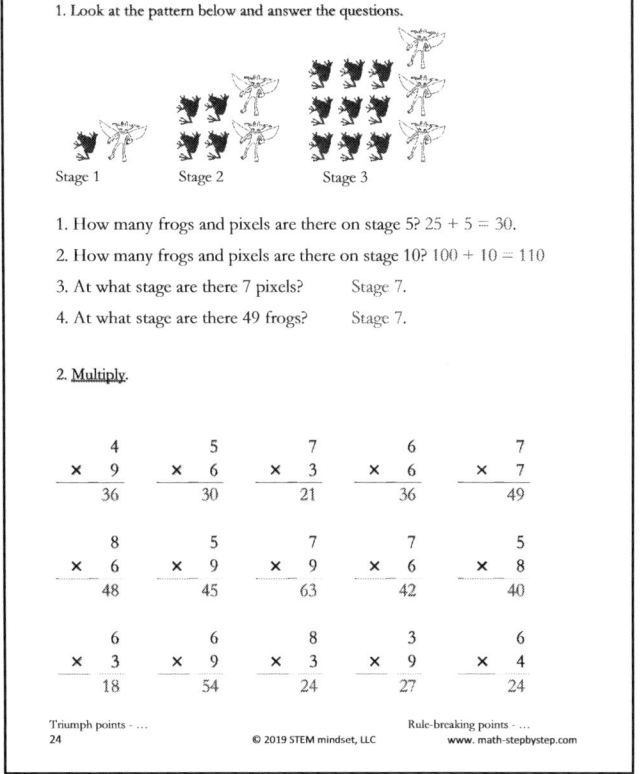

Stage 1   Stage 2   Stage 3

1. How many frogs and pixels are there on stage 5? $25 + 5 = 30$.
2. How many frogs and pixels are there on stage 10? $100 + 10 = 110$
3. At what stage are there 7 pixels?   Stage 7.
4. At what stage are there 49 frogs?   Stage 7.

2. Multiply.

4 × 9 = 36	5 × 6 = 30	7 × 3 = 21	6 × 6 = 36	7 × 7 = 49
8 × 6 = 48	5 × 9 = 45	7 × 9 = 63	7 × 6 = 42	5 × 8 = 40
6 × 3 = 18	6 × 9 = 54	8 × 3 = 24	3 × 9 = 27	6 × 4 = 24

1. <u>Divide</u>.

$$2\overline{)8} = 4 \quad 3\overline{)21} = 7 \quad 4\overline{)36} = 9 \quad 5\overline{)40} = 8 \quad 6\overline{)30} = 5$$

$$8\overline{)56} = 7 \quad 5\overline{)30} = 6 \quad 7\overline{)42} = 6 \quad 6\overline{)36} = 6 \quad 5\overline{)45} = 9$$

$$7\overline{)63} = 9 \quad 8\overline{)72} = 9 \quad 9\overline{)54} = 6 \quad 8\overline{)64} = 8 \quad 9\overline{)81} = 9$$

2. There are five times as many both the tall goal posts and players on broomsticks as flying balls in Quidditch. There are six tall goal posts and fourteen players on broomsticks. How many flying balls are in Quidditch?

Goal posts + Players: 6 + 14 = 20

Flying balls: 20 ÷ 5 = 4

Answer: four flying balls.

---

1. <u>Multiply</u> and <u>find</u> the value. <u>Indicate</u> order of operations.

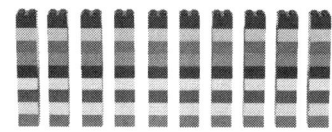

$9 \times 2 = 18 \qquad 9 \times 4 = 36 \qquad 9 \times 6 = 54$

$9 \times 8 = 72 \qquad 9 \times 10 = 90$

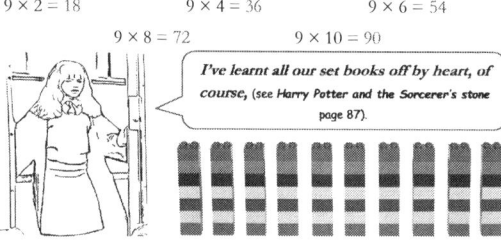

I've learnt all our set books off by heart, of course, (see Harry Potter and the Sorcerer's stone page 87).

$7 \times 2 = 14 \qquad 7 \times 4 = 28 \qquad 7 \times 6 = 42$

$7 \times 8 = 56 \qquad 7 \times 10 = 70$

$\overset{1}{(4 \times 4)} + \overset{4}{(9 \times 2)} + \overset{2}{(7 \times 10)} = 104 \qquad \overset{1}{(6 \times 8)} + \overset{4}{(9 \times 4)} + \overset{2}{(4 \times 8)} = 116$

$(6 \times 10) + (7 \times 6) + (4 \times 2) = 110 \quad (9 \times 6) - (4 \times 10) + (7 \times 4) = 42$

$(9 \times 8) - (4 \times 6) + (6 \times 2) = 60 \quad (9 \times 10) + (4 \times 2) - (7 \times 8) = 42$

$(6 \times 4) - (7 \times 2) + (9 \times 6) = 64 \quad (6 \times 6) + (7 \times 4) - (4 \times 10) = 24$

Dudley, meanwhile, was counting his presents. His face fell. "Thirty-six", (see Harry Potter and the Sorcerer's Stone page 17).

1. <u>Answer</u> the questions.

<u>Imagine</u> that Dudley got three kinds of different birthday presents (video games, racing cars, and remote-control aeroplanes). <u>Find</u> three one-digit numbers of 3 kinds of Dudley's birthday presents.

Note: any number has to be more than 1.

When the sum of these one-digit numbers is 10 and their product is 36: 4+3+3=10; 4×3×3=36.

<u>Draw</u> the dots to show the pattern for the product and <u>circle</u> the bricks to show the number of Dudley's birthday presents.

 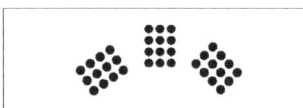

When the sum of these one-digit numbers is 11 and their product is 36: 2+3+6=11; 2×3×6=36.

<u>Draw</u> the dots to show the pattern for the product and <u>circle</u> the bricks to show the number of Dudley's birthday presents.

 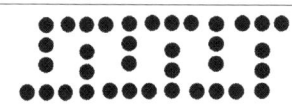

When the sum of these one-digit numbers is 13 and their product is 36: 2+2+9=13; 2×2×9=36.

<u>Draw</u> the dots to show the pattern for the product and <u>circle</u> the bricks to show the number of Dudley's birthday presents.

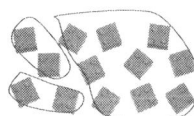 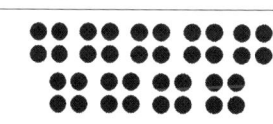

If the final number of Dudley's birthday presents was 39, place the parentheses to make the value correct.

$(5 + 6) \times 4 - 2 = 39$

$47 - 24 \div (6 - 3) = 39$

1. **Multiply.** **Add** the products in each column. The Triwizard Cup goes to the school with the greatest sum.

... become Hogwarts champion...was standing on the grounds... arms raised in triumph in front of the whole school... had just won the Triwizard Tournament (see *Harry Potter and the Goblet of Fire* page 192).

Hogwarts	*Beauxbatons*	*Durmstrang*
9 × 8 = 72	8 × 3 = 24	5 × 7 = 35
9 × 5 = 45	6 × 4 = 24	7 × 3 = 21
7 × 6 = 42	4 × 8 = 32	2 × 7 = 14
7 × 9 = 63	5 × 8 = 40	8 × 7 = 56
9 × 2 = 18	6 × 6 = 36	9 × 9 = 81
8 × 2 = 16	9 × 6 = 54	8 × 8 = 64
7 × 7 = 49	4 × 5 = 20	9 × 4 = 36
6 × 5 = 30	1 × 5 = 5	8 × 5 = 40
335	235	347

---

1. **Answer** the questions. Waaaaaaaaaa!

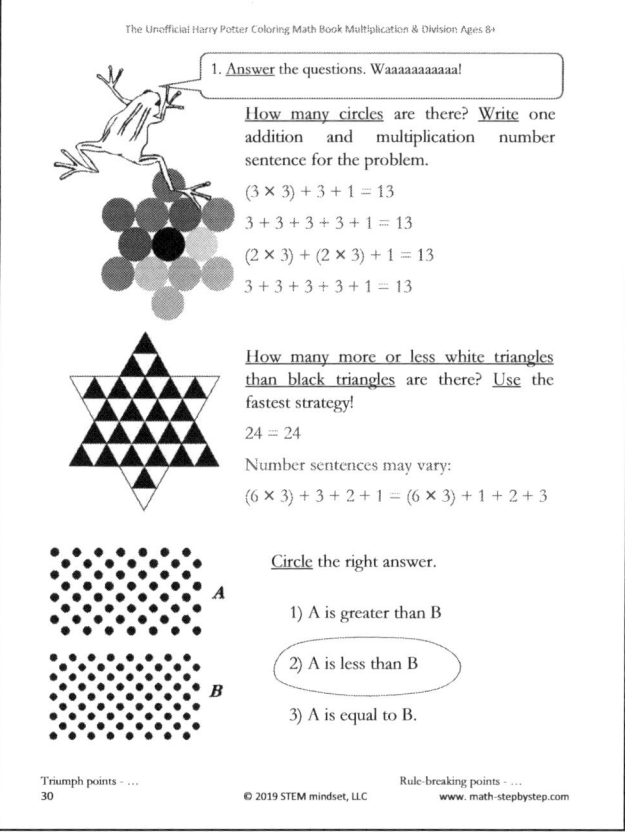

How many circles are there? Write one addition and multiplication number sentence for the problem.

(3 × 4) + 3 + 1 = 13
3 + 3 + 3 + 3 + 1 = 13
(2 × 3) + (2 × 3) + 1 = 13
3 + 3 + 3 + 3 + 1 = 13

How many more or less white triangles than black triangles are there? Use the fastest strategy!

24 = 24

Number sentences may vary:
(6 × 3) + 3 + 2 + 1 = (6 × 3) + 1 + 2 + 3

Circle the right answer.
1) A is greater than B
2) (A is less than B)
3) A is equal to B.

---

1. **Multiply** and **find** the value. **Indicate** the order of operations.

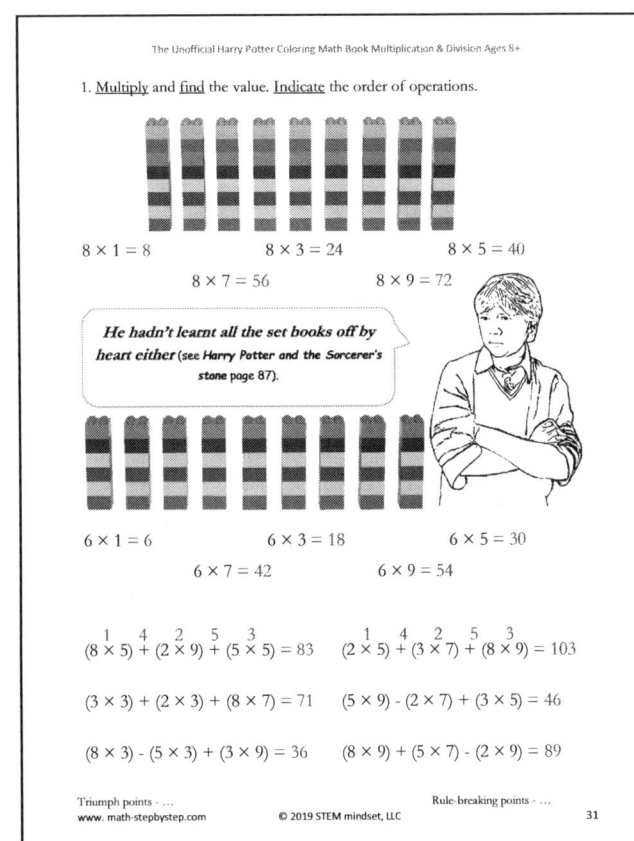

8 × 1 = 8   8 × 3 = 24   8 × 5 = 40
8 × 7 = 56   8 × 9 = 72

He hadn't learnt all the set books off by heart either (see *Harry Potter and the Sorcerer's stone* page 87).

6 × 1 = 6   6 × 3 = 18   6 × 5 = 30
6 × 7 = 42   6 × 9 = 54

$\overset{1}{(8 × 5)} + \overset{4}{(2 × 9)} + \overset{2}{(5 × 5)} = 83$ $\overset{1}{(2 × 5)} + \overset{4}{(3 × 7)} + \overset{2}{(8 × 9)} = 103$

(3 × 3) + (2 × 3) + (8 × 7) = 71   (5 × 9) - (2 × 7) + (3 × 5) = 46

(8 × 3) - (5 × 3) + (3 × 9) = 36   (8 × 9) + (5 × 7) - (2 × 9) = 89

---

1. **Answer** the questions. **Write** one multiplication and addition number sentence for each problem. **Circle** the groups of objects to show your fastest strategy.

How many dots are there?
(3 × 4) + (3 × 4) = 24

How many dots are there?
(6 × 6) + (2 × 5) + 6 = 52
(6 × 5) + (4 × 3) + 10 = 52

One more lesson like that and I might just do a Weasley (see *Harry Potter and the Order of the Phoenix* page 596).

How many dots are there?
(7 × 6) + 3 = 45

How many bronze Knuts are there?
(4 × 5) + (4 × 4) + (4 × 3) + (4 × 2) + (2 × 1) = 58

Page 33

1. <u>Find</u> and <u>circle</u> or <u>cross out</u> the words to find out more about Harry Potter.

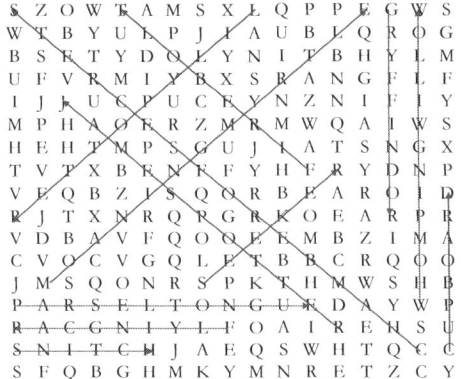

GRYFFINDOR   FIREBOLT
LILY POTTER   JAMES POTTER
PARSELTONGUE   SEEKER
CUPBOARD   SNITCH
WHOMPING WILLOW   FLYING CAR
MOANING MYRTLE   CHAMBER OF SECRETS

Page 34

1. <u>Multiply</u>.

```
   2       5       8       6       6
 × 9     × 4     × 3     × 3     × 7
  18      20      24      18      42

   8       4       7       8       8
 × 5     × 9     × 5     × 6     × 8
  40      36      35      48      64

   6       9       7       5       6
 × 4     × 9     × 3     × 9     × 5
  24      81      21      45      30
```

2. George Weasley collected 72 Chocolate Frog cards with Albus Dumbledore. He collected 8 times as many cards as Fred Weasley. <u>How many cards</u> did Fred collect?

George	George	Fred	George: $72 = 8 \times$ Fred
72	8 × Fred	? 9	Fred $= 72 \div 8 = 9$
			Answer: 9 Chocolate Frog cards.

Page 35

1. <u>Divide</u>.

```
     8        10       10       10       10
  2)16     3)30     4)40     5)50     6)60
    16       30       40       50       60
     0        0        0        0        0

    10        4       10        2        6
  8)80     5)20     7)70     6)12     5)30
    80       20       70       12       30
     0        0        0        0        0

     5        4        5        3        8
  7)35     8)32     9)45     8)24     9)72
    35       32       45       24       72
     0        0        0        0        0
```

2. Ron is thinking of a number. When this number is divided by 6, the answer is 9. <u>What</u> is the number?

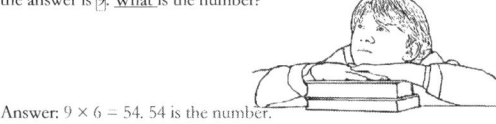

Answer: $9 \times 6 = 54$. 54 is the number.

Page 36

1. <u>Answer</u> the questions.

Don't you care about Gryffindor, do you only care about yourselves, I don't want Slytherin to win the House Cup and you'll lose all the points I got from Professor McGonagall for knowing about Switching Spells (see Harry Potter and the Sorcerer's stone page 127).

Think, Brainer, think! You need more points to win the House Cup!

Find a pair of two numbers: their difference is 12, and the product is 45:
$15 - 3 = 12$; $15 \times 3 = 45$.

Find a pair of two numbers: their difference is 7, and the product is 60:
$12 - 5 = 12$; $12 \times 5 = 60$.

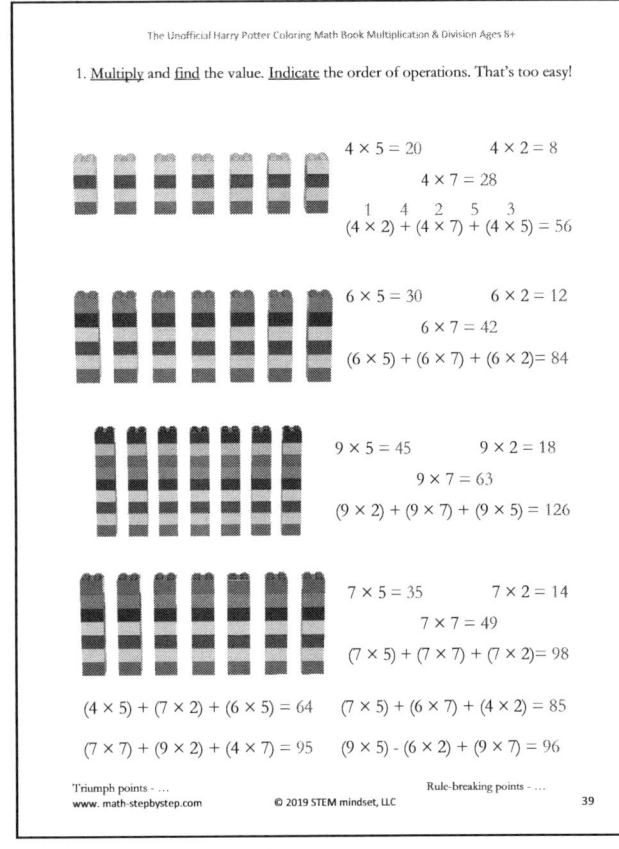
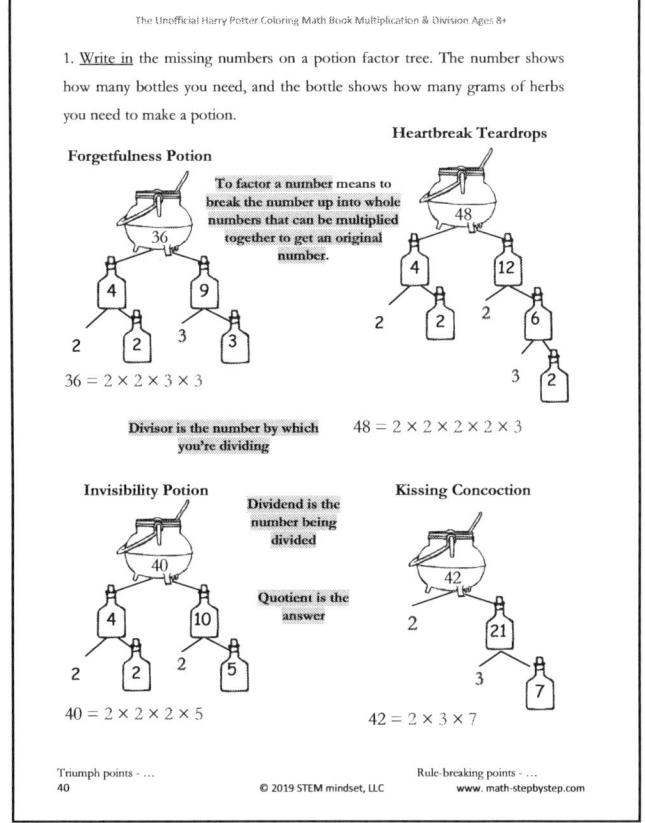

Page 41

1. Circle the right answer.

A 3 × 4 × 7
B 2 × 7 × 6

- A is equal to B
- A is greater than B
- A is less than B

A (8 × 7) - 19
B (9 × 6) - 28

- (A is greater than B)
- A is equal to B
- A is less than B

A (9 × 3) + 14
B (7 × 5) + 28

- A is greater than B
- A is equal to B
- (A is less than B)

A 2 × 2 × 6 × 3
B 4 × 2 × 3 × 2

- (A is greater than B)
- A is less than B
- A is equal to B

Page 42

1. Multiply and find the value. Indicate the order of operations. Ca-ha-ha-ac!

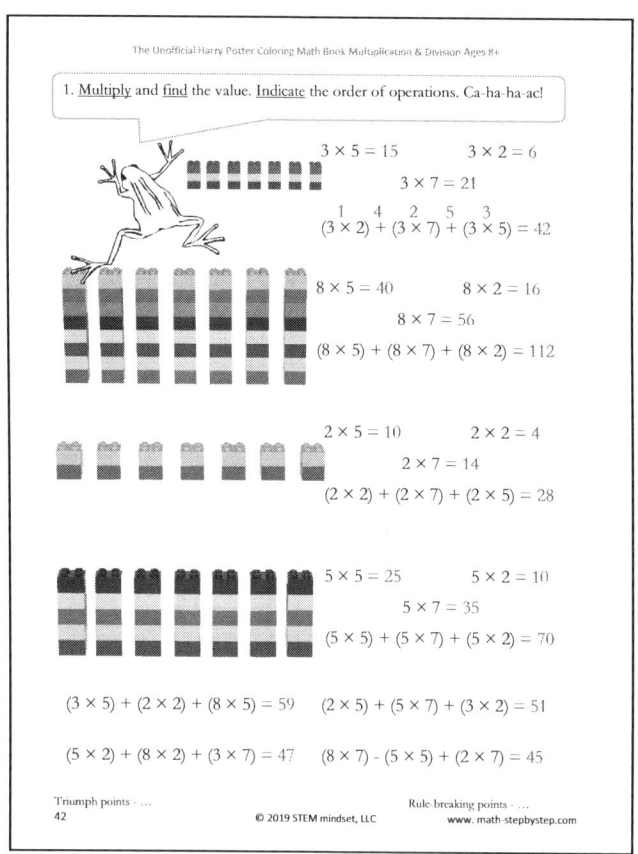

$3 \times 5 = 15 \qquad 3 \times 2 = 6$
$3 \times 7 = 21$
$\overset{1}{(3 \times 2)} + \overset{4}{(3 \times 7)} + \overset{2}{(3 \times 5)} = 42$

$8 \times 5 = 40 \qquad 8 \times 2 = 16$
$8 \times 7 = 56$
$(8 \times 5) + (8 \times 7) + (8 \times 2) = 112$

$2 \times 5 = 10 \qquad 2 \times 2 = 4$
$2 \times 7 = 14$
$\overset{2}{(2 \times 2)} + \overset{5}{(2 \times 7)} + \overset{3}{(2 \times 5)} = 28$

$5 \times 5 = 25 \qquad 5 \times 2 = 10$
$5 \times 7 = 35$
$(5 \times 5) + (5 \times 7) + (5 \times 2) = 70$

$(3 \times 5) + (2 \times 2) + (8 \times 5) = 59 \qquad (2 \times 5) + (5 \times 7) + (3 \times 2) = 51$

$(5 \times 2) + (8 \times 2) + (3 \times 7) = 47 \qquad (8 \times 7) - (5 \times 5) + (2 \times 7) = 45$

Page 43

1. Answer the questions. Write one multiplication and addition, and one division number sentence for each problem. Fill in the missing numbers.

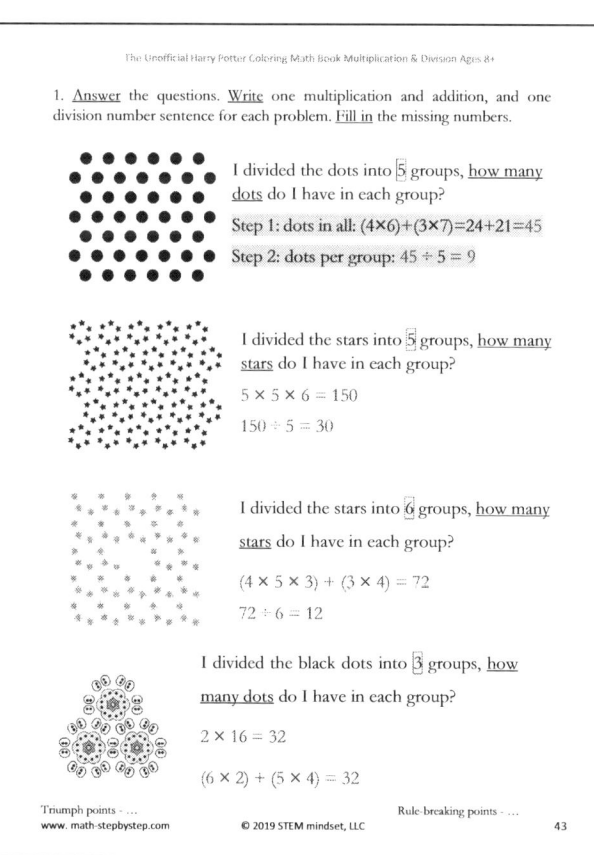

I divided the dots into 5 groups, how many dots do I have in each group?

Step 1: dots in all: (4×6)+(3×7)=24+21=45

Step 2: dots per group: 45 ÷ 5 = 9

I divided the stars into 5 groups, how many stars do I have in each group?

$5 \times 5 \times 6 = 150$
$150 \div 5 = 30$

I divided the stars into 6 groups, how many stars do I have in each group?

$(4 \times 5 \times 3) + (3 \times 4) = 72$
$72 \div 6 = 12$

I divided the black dots into 3 groups, how many dots do I have in each group?

$2 \times 16 = 32$
$(6 \times 2) + (5 \times 4) = 32$

Page 44

1. Find and circle or cross out the words to find out more about Albus Dumbledore.

```
X F B K E D E J F K H N I A G O R O
I T M W E U P A L A J Y D Y V S I Y
N I H I X U K K U Q W L M C C E Z U
E X J S E J Z V Z I H K W I E B N N
O L B D N X H S G X V V E B E I M Q
I K A O L C Y T I L I B I S I V N I
P B D M P I H S D N E I R F Z X C R
E N O M E L T E B R E H S D B C R C
H S Y U V N Q V O V N B U V P O B U
T C H O C O L A T E F R O G C A R D
F K N I T T I N G P A T T E R N S M
O V Z S K L X Z V V I M V K A L U R
R R X W O O L Y S O C K S V Z S V Q
E T B U R N I N G D A Y M Y I T D J
D J M D V W O N Z K D Q Q C Z F D C
R E T S A M D A E H S T R A W G O H
O V G O O S Y N M Q J V F T E E C L
H Q P S S B Q W N W N K J C U D X P
```

HOGWARTS HEADMASTER	SHERBET LEMON
ORDER OF THE PHOENIX	BURNING DAY
CHOCOLATE FROG CARD	MUSIC
FAWKES	FRIENDSHIP
WOOLY SOCKS	INVISIBILITY CLOAK
KNITTING PATTERNS	WISDOM

Page 45

1. <u>Multiply</u>.

```
   2        5        7        6        9
 × 5      × 8      × 3      × 6      × 7
 ---      ---      ---      ---      ---
  10       40       21       36       63

   8        6        7        4        7
 × 4      × 9      × 4      × 6      × 8
 ---      ---      ---      ---      ---
  32       54       28       24       56

   4        9        7        8        6
 × 4      × 6      × 5      × 9      × 3
 ---      ---      ---      ---      ---
  16       54       35       72       18
```

2. There are 6 trees along the road in front of Hagrid's house. The trees are 2 ft away from each other. <u>How far (in inches)</u> is the third tree from the first tree?

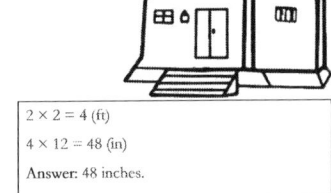

Trees	Ft between	Intervals	Inches
3	2	2	24

$2 \times 2 = 4$ (ft)
$4 \times 12 = 48$ (in)
Answer: 48 inches.

Page 46

1. Mr. Dursley saw many people whispering excitedly and dressed in cloaks that day. The number of people he met in the morning was four times the number of people he met during the lunch. There were 40 people in cloaks altogether. <u>How many more or less people in cloaks in the morning than in the afternoon</u> were there?

Morning	Lunch	In all	Morning > or < Lunch
4 × L, 32	? L; 8	40	24 more

Morning + Lunch = 40
4×L + L = 40
5 × L = 40
L = 8   M = 4 × 8   M = 32
Answer: 32 − 8 = 24 (more).

2. <u>Write in</u> the missing numbers on a potion factor tree. The number shows how many bottles you need, and the bottle shows how many grams of herbs you need to make a potion.

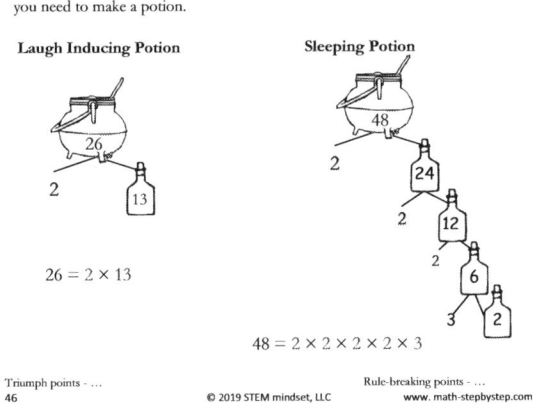

Laugh Inducing Potion

$26 = 2 \times 13$

Sleeping Potion

$48 = 2 \times 2 \times 2 \times 2 \times 3$

Page 47

1. <u>Divide</u>.

```
    9        8        9        6        9
 2)18     3)24     4)36     5)30     6)54
   18       24       36       30       54
   --       --       --       --       --
    0        0        0        0        0

    5        9        3        7       10
 8)40     5)45     7)21     6)42     5)50
   40       45       21       42       50
   --       --       --       --       --
    0        0        0        0        0

    7        7        7        2        3
 7)49     8)56     9)63     8)16     9)27
   49       56       63       16       27
   --       --       --       --       --
    0        0        0        0        0
```

2. Mr. Dursley counted 21 owls in the morning. There were two times as many flying owls as sitting owls on the trees. <u>How many owls</u> were flying?

2 × Sitting + Sitting = 21
21 ÷ 3 = 7 (Sitting)
2 × 7 = 14 (Flying)

Page 48

Dobby is a free elf, sir, and Dobby gets a Galleon a week and one day off a month! (see Harry Potter and the Goblet of Fire page 379).

1. <u>How much</u> will Dobby earn a year?

53 weeks: 53 × 1 = 53 (dollars).

Very fortunate... You've lost me my servant, boy! (see Harry Potter and the Chamber of Secrets pages 336-338).

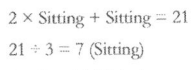

Page 49

1. <u>Multiply</u> and <u>find</u> the value. <u>Indicate</u> the order of operations.

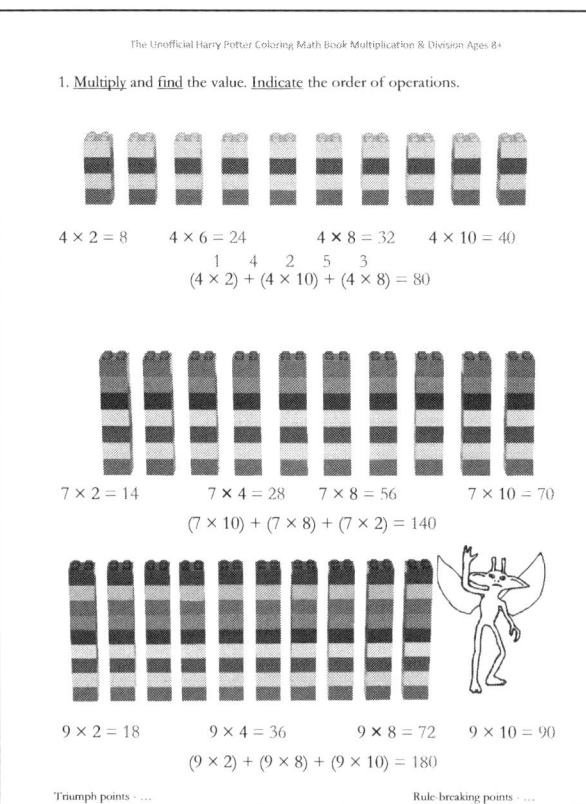

$4 \times 2 = 8$ $4 \times 6 = 24$ $4 \times 8 = 32$ $4 \times 10 = 40$

$\quad\quad\quad\quad 1 \quad\quad 4 \quad 2 \quad\quad 5 \quad\quad 3$
$(4 \times 2) + (4 \times 10) + (4 \times 8) = 80$

$7 \times 2 = 14$ $7 \times 4 = 28$ $7 \times 8 = 56$ $7 \times 10 = 70$

$(7 \times 10) + (7 \times 8) + (7 \times 2) = 140$

$9 \times 2 = 18$ $9 \times 4 = 36$ $9 \times 8 = 72$ $9 \times 10 = 90$

$(9 \times 2) + (9 \times 8) + (9 \times 10) = 180$

Page 50

1. <u>Multiply</u>. <u>Use</u> five multiplication number sentences for each problem (decompose the factors). The first one is done for you.

$8 \times 8 = 4 \times 2 \times 4 \times 2 = 1 \times 8 \times 8 \times 1 = 2 \times 2 \times 2 \times 2 \times 2 \times 2 =$
$= 16 \times 4 = 64$

$9 \times 8 = 3 \times 3 \times 4 \times 2 = 1 \times 9 \times 8 \times 1 = 3 \times 3 \times 2 \times 2 \times 2 =$
$= 3 \times 24 = 4 \times 18 = 72$

$10 \times 6 = 2 \times 5 \times 3 \times 2 = 1 \times 10 \times 6 \times 1 = 2 \times 15 \times 2 = 3 \times 20 =$
$= 4 \times 15 = 5 \times 12 = 60$

$8 \times 6 = 2 \times 3 \times 4 \times 2 = 1 \times 6 \times 8 \times 1 = 2 \times 3 \times 2 \times 2 \times 2 =$
$= 4 \times 12 = 2 \times 24 = 48$

2. <u>Help</u> me get the Triwizard Cup. Don't leave me alone! Help needed!

Page 51

1. <u>Multiply</u> and <u>find</u> the value. <u>Indicate</u> the order of operations.

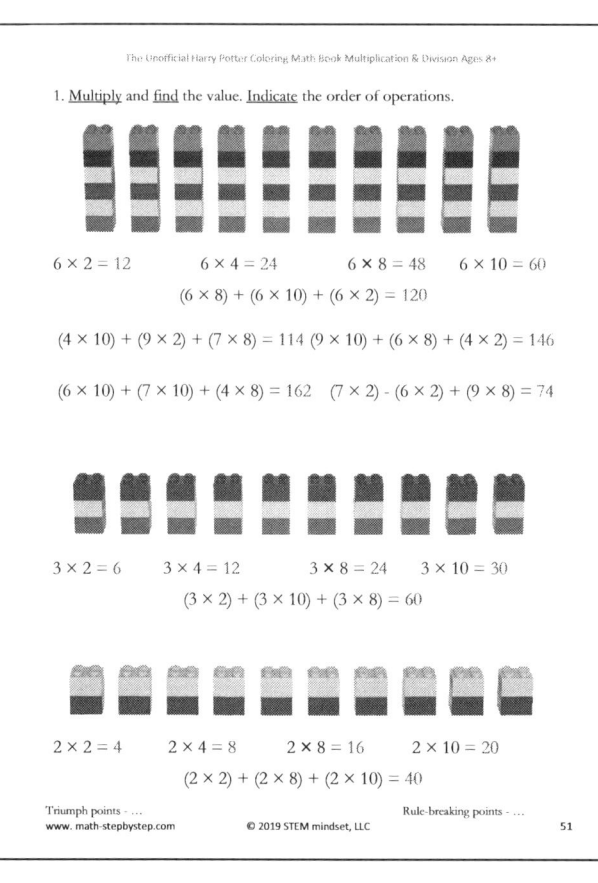

$6 \times 2 = 12$ $6 \times 4 = 24$ $6 \times 8 = 48$ $6 \times 10 = 60$

$(6 \times 8) + (6 \times 10) + (6 \times 2) = 120$

$(4 \times 10) + (9 \times 2) + (7 \times 8) = 114 \quad (9 \times 10) + (6 \times 8) + (4 \times 2) = 146$

$(6 \times 10) + (7 \times 10) + (4 \times 8) = 162 \quad (7 \times 2) - (6 \times 2) + (9 \times 8) = 74$

$3 \times 2 = 6$ $3 \times 4 = 12$ $3 \times 8 = 24$ $3 \times 10 = 30$

$(3 \times 2) + (3 \times 10) + (3 \times 8) = 60$

$2 \times 2 = 4$ $2 \times 4 = 8$ $2 \times 8 = 16$ $2 \times 10 = 20$

$(2 \times 2) + (2 \times 8) + (2 \times 10) = 40$

Page 52

1. <u>Answer</u> the questions. <u>Write</u> one multiplication and addition, and one division number sentence for each problem.

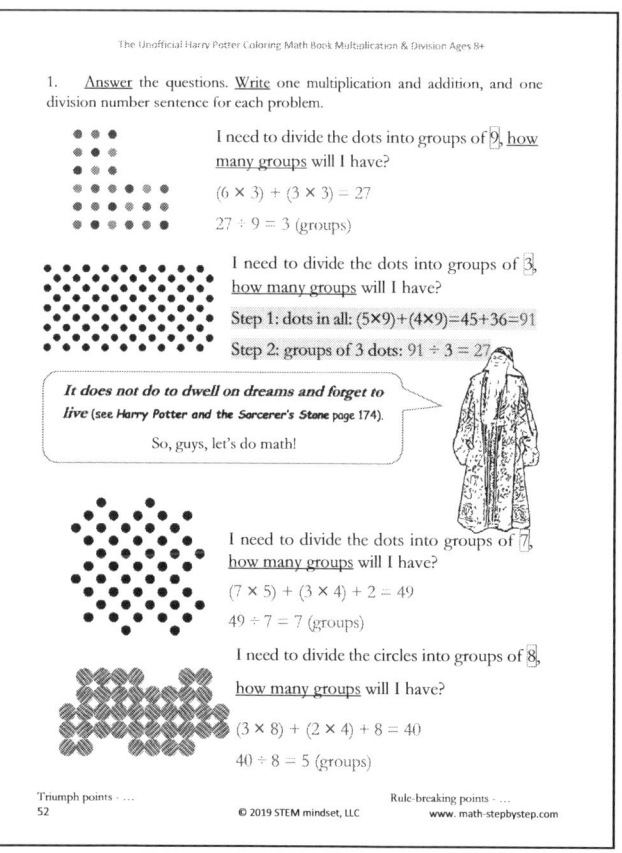

I need to divide the dots into groups of 9, how many groups will I have?

$(6 \times 3) + (3 \times 3) = 27$

$27 \div 9 = 3$ (groups)

I need to divide the dots into groups of 3, how many groups will I have?

Step 1: dots in all: $(5 \times 9) + (4 \times 9) = 45 + 36 = 91$

Step 2: groups of 3 dots: $91 \div 3 = 27$

It does not do to dwell on dreams and forget to live (see *Harry Potter and the Sorcerer's Stone* page 174).

So, guys, let's do math!

I need to divide the dots into groups of 7, how many groups will I have?

$(7 \times 5) + (3 \times 4) + 2 = 49$

$49 \div 7 = 7$ (groups)

I need to divide the circles into groups of 8, how many groups will I have?

$(3 \times 8) + (2 \times 4) + 8 = 40$

$40 \div 8 = 5$ (groups)

1. <u>Write in</u> the missing numbers on a potion factor tree. The number shows how many bottles you need, and the bottle shows how many grams of herbs you need to make a potion.

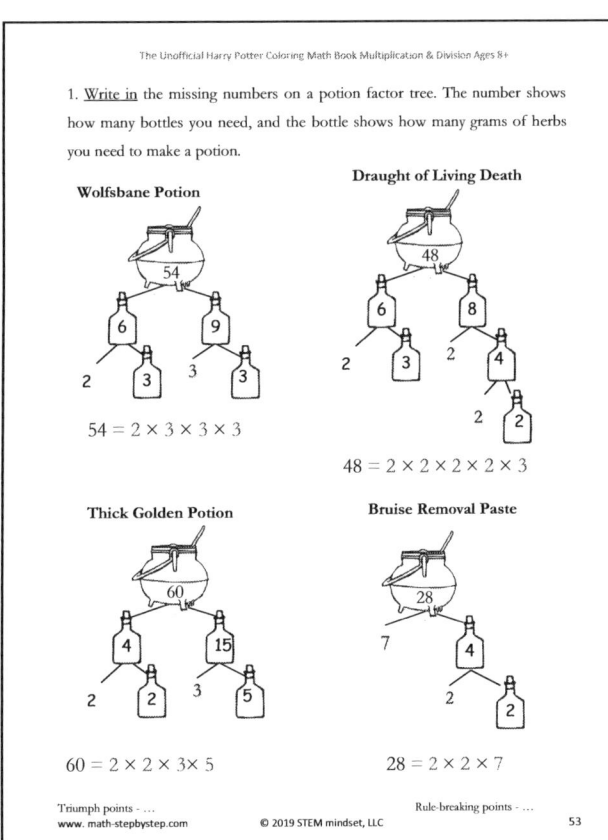

$54 = 2 \times 3 \times 3 \times 3$

$48 = 2 \times 2 \times 2 \times 2 \times 3$

$60 = 2 \times 2 \times 3 \times 5$

$28 = 2 \times 2 \times 7$

---

1. <u>Multiply</u> and <u>find</u> the value. <u>Indicate</u> the order of operations.

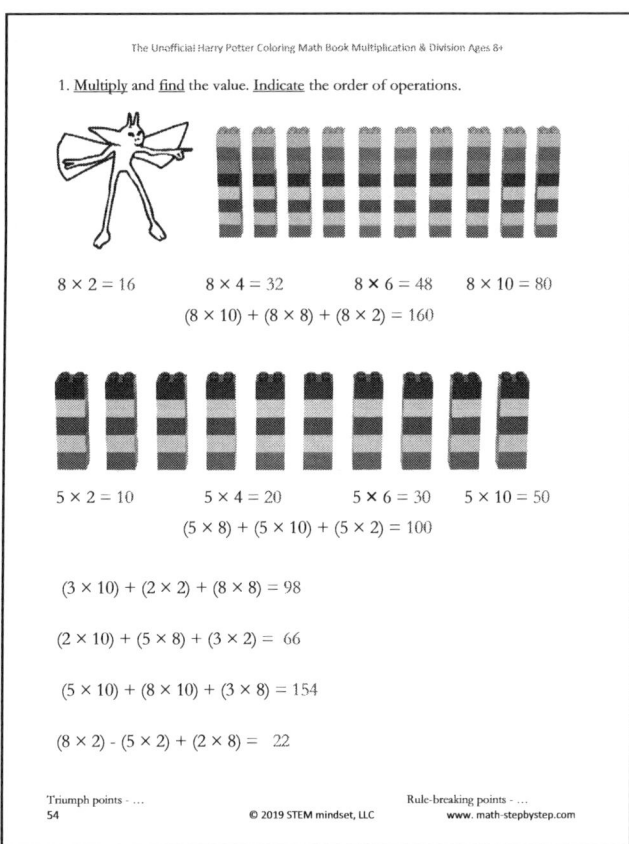

$8 \times 2 = 16 \qquad 8 \times 4 = 32 \qquad 8 \times 6 = 48 \qquad 8 \times 10 = 80$

$(8 \times 10) + (8 \times 8) + (8 \times 2) = 160$

$5 \times 2 = 10 \qquad 5 \times 4 = 20 \qquad 5 \times 6 = 30 \qquad 5 \times 10 = 50$

$(5 \times 8) + (5 \times 10) + (5 \times 2) = 100$

$(3 \times 10) + (2 \times 2) + (8 \times 8) = 98$

$(2 \times 10) + (5 \times 8) + (3 \times 2) = 66$

$(5 \times 10) + (8 \times 10) + (3 \times 8) = 154$

$(8 \times 2) - (5 \times 2) + (2 \times 8) = 22$

He...didn't dare mention platform nine and three-quarters (see Harry Potter and the Sorcerer's Stone page 76).

Professor Dumbledore offered Dobby ten Galleons a week ... but Dobby beat him down... (see Harry Potter and the Goblet of Fire page 379).

1. <u>How much more would Dobby earn a year if he accepted Professor Dumbledore's offer?</u> <u>Compare</u> with p. 48.

$53 \times 10 = 530$ (Galleons if he earns 10 Galleons a week)

$530 - 53 = 477$ (Galleons Dobby would have earned if he had accepted Professor Dumbledore's offer).

---

1. <u>Find</u> and <u>circle</u> or <u>cross out</u> the words to find out more about Fawkes.

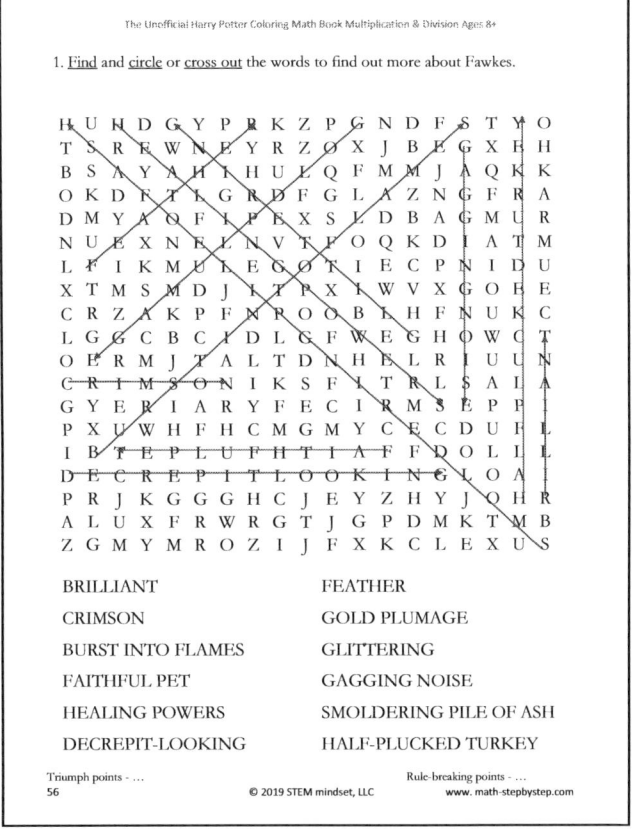

BRILLIANT — FEATHER
CRIMSON — GOLD PLUMAGE
BURST INTO FLAMES — GLITTERING
FAITHFUL PET — GAGGING NOISE
HEALING POWERS — SMOLDERING PILE OF ASH
DECREPIT-LOOKING — HALF-PLUCKED TURKEY

1. Multiply.

7 × 5 = 35	8 × 8 = 64	7 × 6 = 42	9 × 6 = 54	9 × 3 = 27
8 × 5 = 40	4 × 9 = 36	7 × 7 = 49	6 × 6 = 36	4 × 4 = 16
4 × 5 = 20	9 × 7 = 63	5 × 5 = 25	2 × 9 = 18	6 × 4 = 24

2. There are 8 lamp posts along the road of Privet Drive. The lamp posts are 7 ft apart from each other. How far (in inches) is the fifth lamp post from the first one?

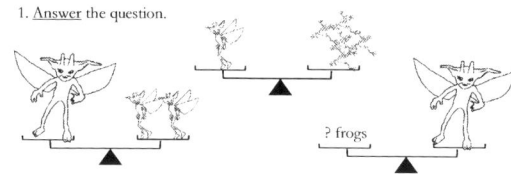

Lamp posts	Ft between	Intervals	Inches
5	7	4	336

4 × 7 = 28 (ft is the fifth lamp post)
28 × 12 = 280 + 56 = 336 (inches)

Answer: 336 inches.

---

1. Answer the question.

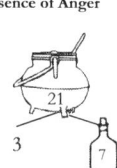 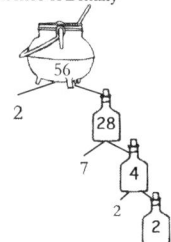

? frogs

1 big pixie = 2 small pixies   1 small pixie = 5 frogs
1 big pixie = 2 small pixies × 5 frogs = 10 frogs
Answer: 10 frogs.

2. Write in the missing numbers on a potion factor tree. The number shows how many bottles you need, and the bottle shows how many grams of herbs you need to make a potion.

Essence of Anger

21
2   3
     7

21 = 3 × 7

Essence of Dittany

56
2   28
    7   4
         2   2

56 = 2 × 2 × 2 × 7

---

1. Divide.

2)4 = 2, 4, 0	3)9 = 3, 9, 0	4)12 = 3, 12, 0	5)10 = 2, 10, 0	6)12 = 2, 12, 0
8)16 = 2, 16, 0	5)15 = 3, 15, 0	7)14 = 2, 14, 0	6)18 = 3, 18, 0	5)25 = 5, 25, 0
7)28 = 4, 28, 0	8)24 = 3, 24, 0	9)27 = 3, 27, 0	8)32 = 4, 32, 0	9)18 = 2, 18, 0

2. I'm thinking of a number. When I multiply 9 to the number, the answer is 36. What is the number?

The unknown number I write as x
Then, 9 × x = 36 →
How to find x? x = 36 ÷ 9 = 4
Answer: The number is 4.

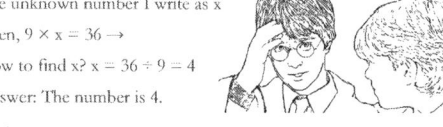

---

1. Find the multiplier. Hint: the multiplier tells you how many equal groups you have put together.

2 times 2 =	4	3 times 2 is	6
4 times 2 =	8	5 times 2 is	10
6 times 2 =	12	7 times 2 is	14
8 times 2 =	16	9 times 2 is	18
3 times 3 =	9	4 times 4 is	16
5 times 5 =	25	6 times 6 is	36
7 times 7 =	49	8 times 8 is	64
9 times 9 =	81	3 times 4 is	12
3 times 5 =	15	3 times 6 is	18
3 times 7 =	21	3 times 8 is	24
3 times 9 =	27	4 times 5 is	20
4 times 6 =	24	4 times 7 is	28
4 times 8 =	32	4 times 9 is	36
5 times 6 =	30	5 times 7 is	35
5 times 8 =	40	5 times 9 is	45
6 times 7 =	42	6 times 8 is	48
6 times 9 =	54	7 times 8 is	56
7 times 9 =	63	8 times 9 is	72

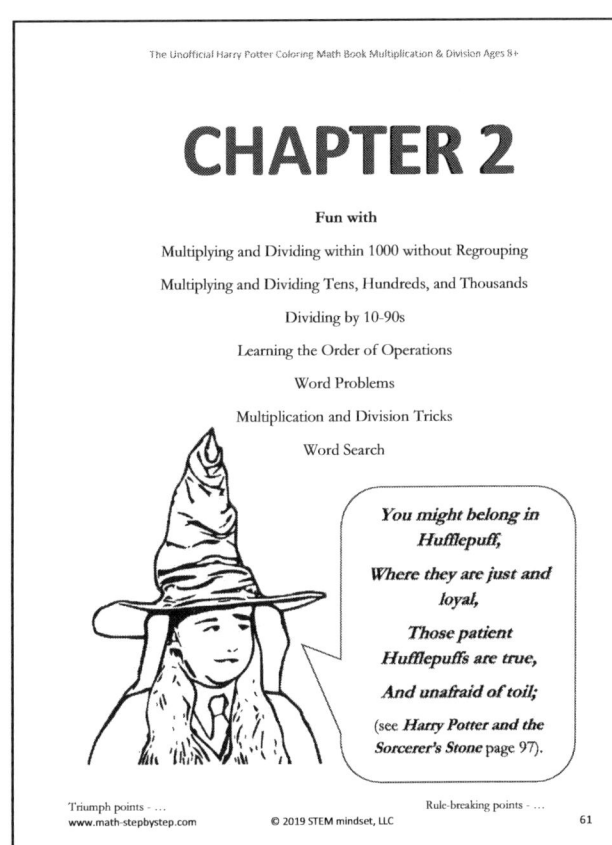

CHAPTER 2

Fun with

Multiplying and Dividing within 1000 without Regrouping

Multiplying and Dividing Tens, Hundreds, and Thousands

Dividing by 10-90s

Learning the Order of Operations

Word Problems

Multiplication and Division Tricks

Word Search

You might belong in Hufflepuff,

Where they are just and loyal,

Those patient Hufflepuffs are true,

And unafraid of toil;

(see *Harry Potter and the Sorcerer's Stone* page 97).

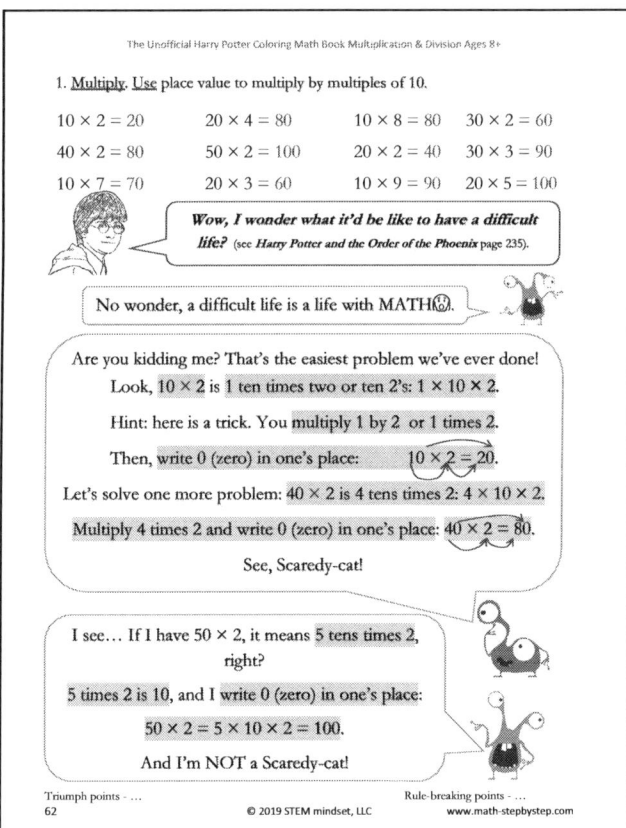

1. <u>Multiply</u>. <u>Use</u> place value to multiply by multiples of 10.

$10 \times 2 = 20$	$20 \times 4 = 80$	$10 \times 8 = 80$	$30 \times 2 = 60$
$40 \times 2 = 80$	$50 \times 2 = 100$	$20 \times 2 = 40$	$30 \times 3 = 90$
$10 \times 7 = 70$	$20 \times 3 = 60$	$10 \times 9 = 90$	$20 \times 5 = 100$

Wow, I wonder what it'd be like to have a difficult life? (see *Harry Potter and the Order of the Phoenix* page 235).

No wonder, a difficult life is a life with MATH☺.

Are you kidding me? That's the easiest problem we've ever done! Look, 10×2 is 1 ten times two or ten 2's: $1 \times 10 \times 2$.

Hint: here is a trick. You multiply 1 by 2 or 1 times 2.

Then, write 0 (zero) in one's place: $10 \times 2 = 20$.

Let's solve one more problem: 40×2 is 4 tens times 2: $4 \times 10 \times 2$.

Multiply 4 times 2 and write 0 (zero) in one's place: $40 \times 2 = 80$.

See, Scaredy-cat!

I see... If I have 50×2, it means 5 tens times 2, right?

5 times 2 is 10, and I write 0 (zero) in one's place: $50 \times 2 = 5 \times 10 \times 2 = 100$.

And I'm NOT a Scaredy-cat!

...I want to fix that in my memory forever, (see *Harry Potter and the Goblet of Fire* page 207).

No, I don't want to fix this trick in my memory forever – multiply here, write zero there!

I'm going to keep going until I succeed – or I die (see *Harry Potter and the Deathly Hallows* page 569).

No drama, guys, we've gotta make it out alive! You multiply each number starting from ones, then tens, then hundreds, and so on:

```
   2 0
 ×   4
 ─────
   8 0
```

4 times 0 equals 0, and you write 0 in one's place: $20 \times 4 = \ldots\, 0$.

Then, 4 times 2 equals 8 and you write 8 in ten's place: $20 \times 4 = 8\,0$.

WRONG:
```
     4
 × 2 0
 ─────
   8 0
```

Hint: use the small number for the multiplier!

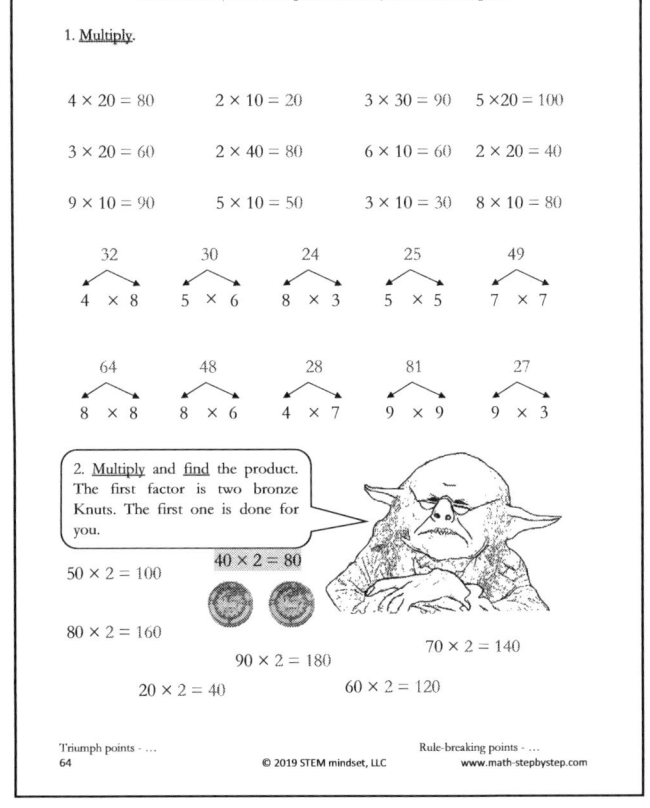

1. <u>Multiply</u>.

$4 \times 20 = 80$	$2 \times 10 = 20$	$3 \times 30 = 90$	$5 \times 20 = 100$
$3 \times 20 = 60$	$2 \times 40 = 80$	$6 \times 10 = 60$	$2 \times 20 = 40$
$9 \times 10 = 90$	$5 \times 10 = 50$	$3 \times 10 = 30$	$8 \times 10 = 80$

32	30	24	25	49
4 × 8	5 × 6	8 × 3	5 × 5	7 × 7

64	48	28	81	27
8 × 8	8 × 6	4 × 7	9 × 9	9 × 3

2. <u>Multiply</u> and <u>find</u> the product. The first factor is two bronze Knuts. The first one is done for you.

$50 \times 2 = 100$ $40 \times 2 = 80$

$80 \times 2 = 160$

$90 \times 2 = 180$ $70 \times 2 = 140$

$20 \times 2 = 40$ $60 \times 2 = 120$

Hope to see you in Hufflepuff!... My old House, you know (see *Harry Potter and the Sorcerer's Stone* page 96).

Well, I like this House, too.

1. There are 27 pixies in all. I divided the pixies into groups of 3. How many pixies are big?

(27 ÷ 3) × 2 = 18 (big pixies)

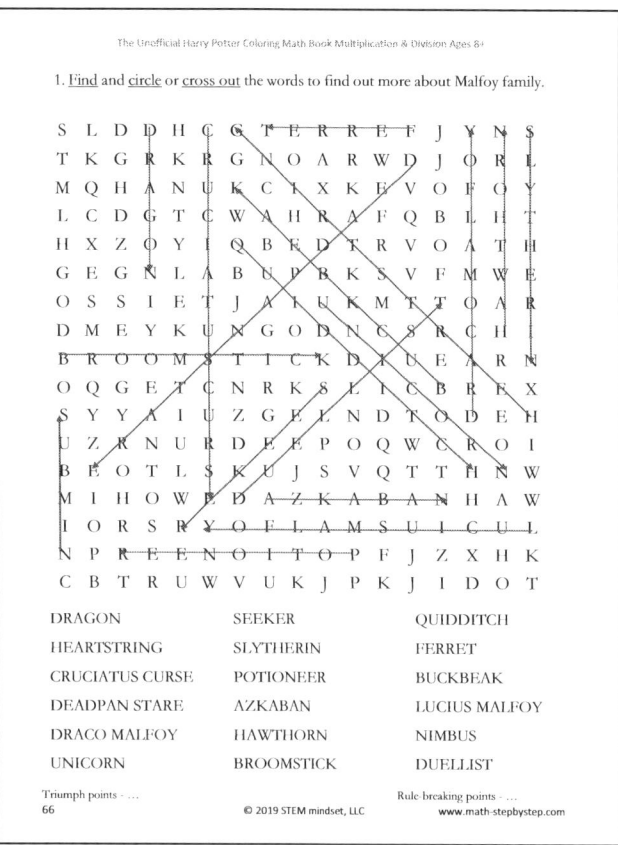

1. Find and circle or cross out the words to find out more about Malfoy family.

DRAGON	SEEKER	QUIDDITCH
HEARTSTRING	SLYTHERIN	FERRET
CRUCIATUS CURSE	POTIONEER	BUCKBEAK
DEADPAN STARE	AZKABAN	LUCIUS MALFOY
DRACO MALFOY	HAWTHORN	NIMBUS
UNICORN	BROOMSTICK	DUELLIST

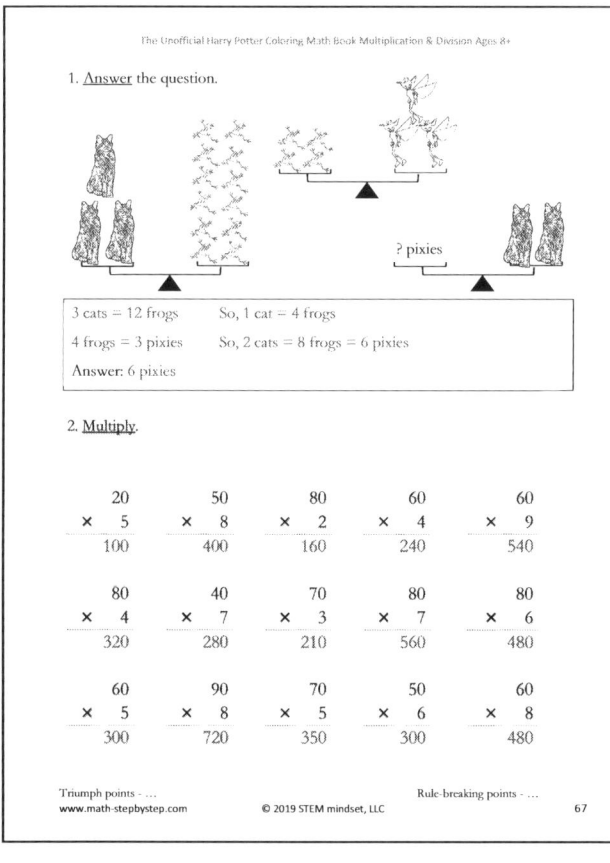

1. Answer the question.

3 cats = 12 frogs   So, 1 cat = 4 frogs
4 frogs = 3 pixies So, 2 cats = 8 frogs = 6 pixies
Answer: 6 pixies

2. Multiply.

20	50	80	60	60
× 5	× 8	× 2	× 4	× 9
100	400	160	240	540

80	40	70	80	80
× 4	× 7	× 3	× 7	× 6
320	280	210	560	480

60	90	70	50	60
× 5	× 8	× 5	× 6	× 8
300	720	350	300	480

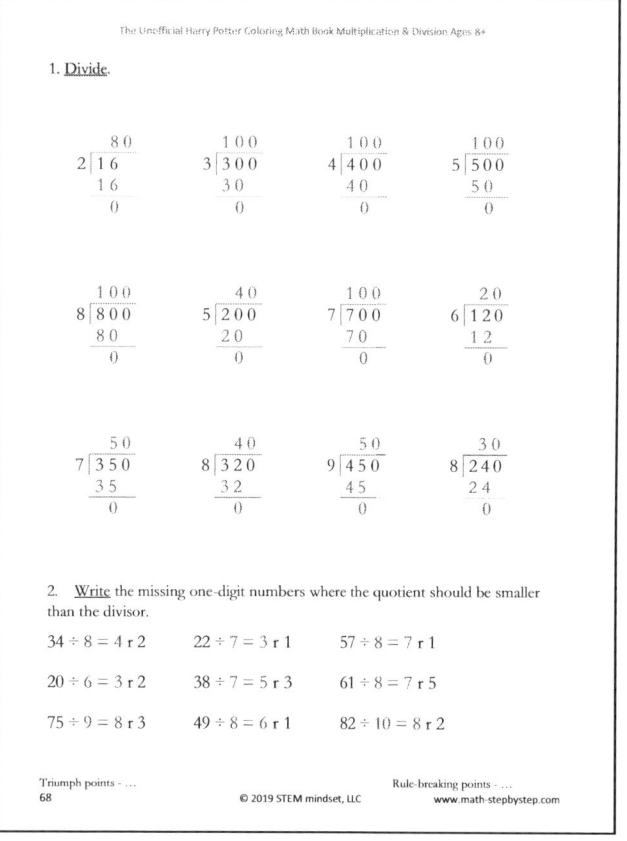

1. Divide.

2)16 = 80   3)300 = 100   4)400 = 100   5)500 = 100
8)800 = 100 5)200 = 40   7)700 = 100   6)120 = 20
7)350 = 50   8)320 = 40   9)450 = 50   8)240 = 30

2. Write the missing one-digit numbers where the quotient should be smaller than the divisor.

34 ÷ 8 = 4 r 2   22 ÷ 7 = 3 r 1   57 ÷ 8 = 7 r 1
20 ÷ 6 = 3 r 2   38 ÷ 7 = 5 r 3   61 ÷ 8 = 7 r 5
75 ÷ 9 = 8 r 3   49 ÷ 8 = 6 r 1   82 ÷ 10 = 8 r 2

1. <u>Write in</u> the missing numbers on a potion factor tree. The number shows how many bottles you need, and the bottle shows how many grams of herbs you need to make a potion.

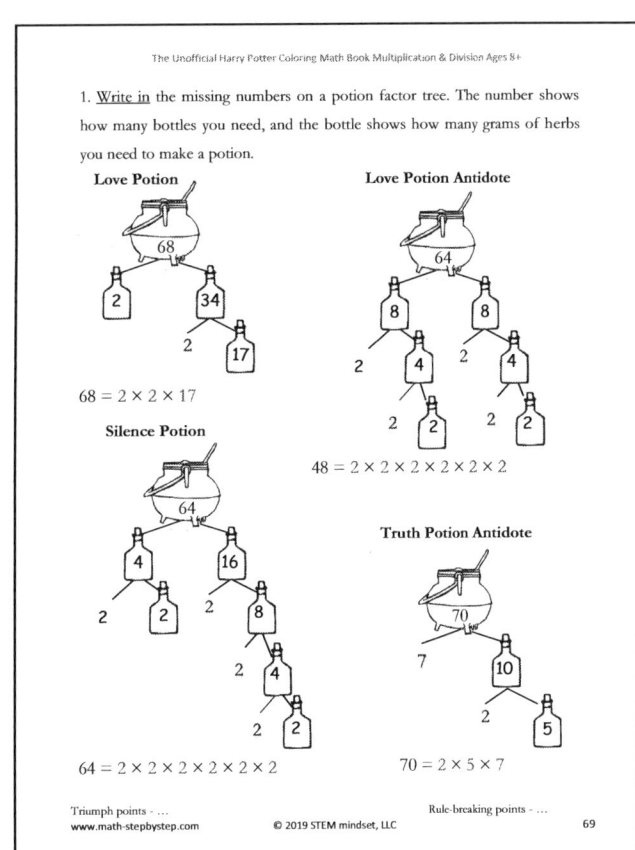

$68 = 2 \times 2 \times 17$

$48 = 2 \times 2 \times 2 \times 2 \times 2$

$64 = 2 \times 2 \times 2 \times 2 \times 2 \times 2$

$70 = 2 \times 5 \times 7$

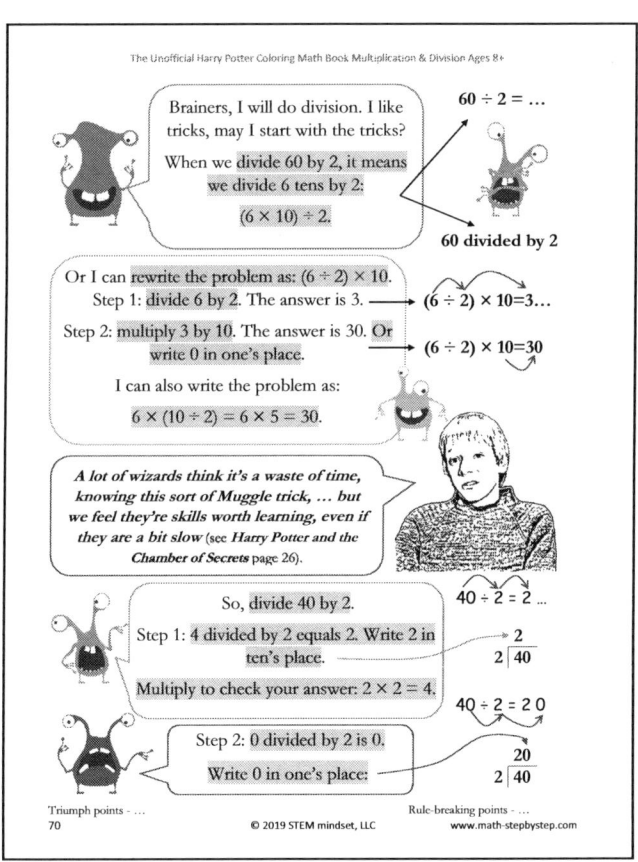

Brainers, I will do division. I like tricks, may I start with the tricks?

When we <u>divide 60 by 2</u>, it means we divide 6 tens by 2: $(6 \times 10) \div 2$.

$60 \div 2 = \ldots$

60 divided by 2

Or I can <u>rewrite</u> the problem as: $(6 \div 2) \times 10$.
Step 1: divide 6 by 2. The answer is 3. → $(6 \div 2) \times 10 = 3\ldots$
Step 2: multiply 3 by 10. The answer is 30. Or write 0 in one's place. → $(6 \div 2) \times 10 = 30$

I can also write the problem as:
$6 \times (10 \div 2) = 6 \times 5 = 30$.

A lot of wizards think it's a waste of time, knowing this sort of Muggle trick, … but we feel they're skills worth learning, even if they are a bit slow (see *Harry Potter and the Chamber of Secrets* page 26).

So, divide 40 by 2.
Step 1: 4 divided by 2 equals 2. Write 2 in ten's place.
Multiply to check your answer: $2 \times 2 = 4$.

$40 \div 2 = 2\ldots$

2
$2\overline{)40}$

$40 \div 2 = 20$

Step 2: 0 divided by 2 is 0. Write 0 in one's place:

20
$2\overline{)40}$

1. <u>Divide</u>.

$40 \div 2 = 20$ $80 \div 4 = 20$ $60 \div 2 = 30$ $90 \div 3 = 30$

$70 \div 7 = 10$ $80 \div 2 = 40$ $60 \div 3 = 20$ $80 \div 8 = 10$

$50 \div 5 = 10$ $90 \div 9 = 10$ $30 \div 3 = 10$ $100 \div 5 = 20$

$40 \div 4 = 10$ $80 \div 8 = 10$ $60 \div 6 = 10$ $100 \div 2 = 50$

10: 2×5 30: 2×15 60: 2×30 70: 2×35 20: 2×10

90: 2×45 80: 2×40 100: 2×50 40: 2×20 50: 2×25

2. <u>Divide</u> and <u>find</u> the quotient. The divisor is two silver Sickles. The first task is done for you.

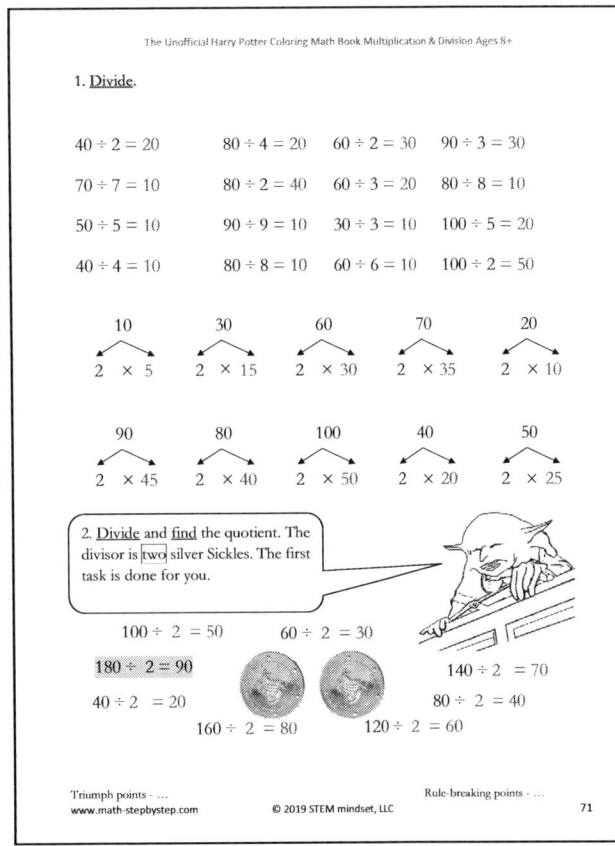

$100 \div 2 = 50$ $60 \div 2 = 30$
$180 \div 2 = 90$
$40 \div 2 = 20$ $140 \div 2 = 70$
$160 \div 2 = 80$ $80 \div 2 = 40$
$120 \div 2 = 60$

1. <u>Multiply</u>. <u>Use</u> five multiplication ways for each problem. The first one is done for you.

$12 \times 4 = 4 \times 3 \times 2 \times 2 = 2 \times 6 \times 4 \times 1 =$
$= 6 \times 8 = 24 \times 2 = 2 \times 2 \times 3 \times 2 \times 2$
$= 2 \times 2 \times 3 \times 2 \times 2 = 48$

$14 \times 6 = 2 \times 7 \times 3 \times 2 = 1 \times 14 \times 3 \times 2 = 2 \times 21 \times 2 = 4 \times 21 =$
$= 3 \times 4 \times 7 = 7 \times 12 = 84$

$16 \times 4 = 4 \times 4 \times 2 \times 2 = 2 \times 8 \times 2 \times 2 = 2 \times 2 \times 2 \times 2 \times 2 \times 2 =$
$= 4 \times 8 \times 2 = 2 \times 32 = 64$

$15 \times 6 = 5 \times 3 \times 3 \times 2 = 1 \times 9 \times 10 \times 1 = 2 \times 45 = 3 \times 15 \times 2 =$
$= 30 \times 3 = 90$

2. <u>Help</u> the Brainer get the Triwizard Cup.

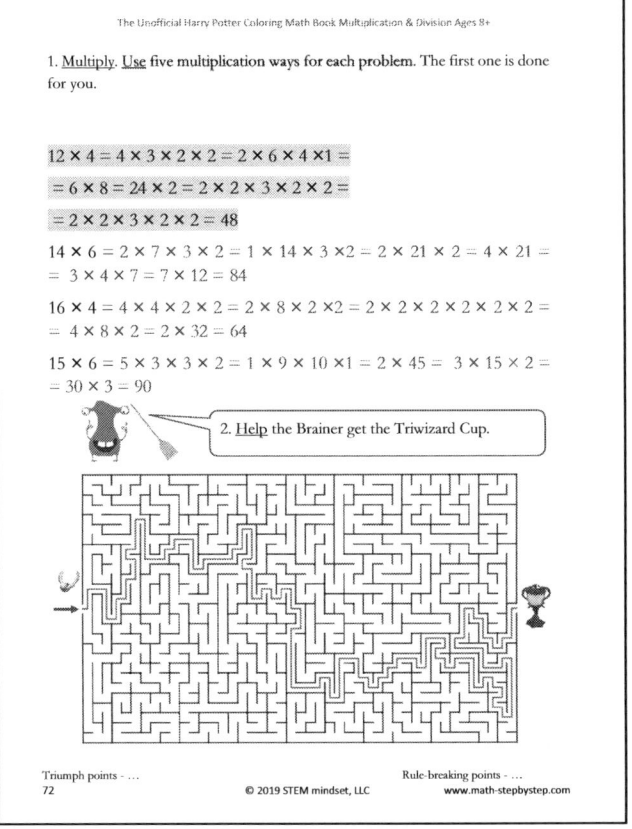

1. Harry spent 33 silver Sickles on Chocolate Frogs, Pumpkin Pasties, and Licorice Wands. Chocolate Frogs cost 6 silver Sickles, and Pumpkin Pasties cost four times as much as Chocolate Frogs. How much more or less did he spend on Chocolate Frogs than on Licorice Wands?

Spent altogether	Frogs	Pasties	Wands	Frogs > or < Wands
33	6	4 × 6 ? 24	? 3	? 3

4 × 6 = 24 (pasties)
33 − 30 = 3 (wands)
6 − 3 = 3 (more)

Answer: 3 more.

2. Write in the missing numbers on a potion factor tree. The number shows how many bottles you need, and the bottle shows how many grams of herbs you need to make a potion.

Essence of Happiness

Essence of Sadness

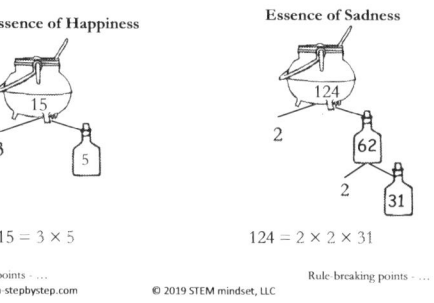

15 = 3 × 5

124 = 2 × 2 × 31

Triumph points - …   Rule-breaking points - …
www.math-stepbystep.com   © 2019 STEM mindset, LLC   73

---

1. Multiply.

18 → 2 × 9   27 → 3 × 7   18 → 6 × 3   40 → 5 × 8   12 → 2 × 6

35 → 7 × 5   56 → 7 × 8   42 → 6 × 7   64 → 8 × 8   36 → 4 × 9

2. Divide.

20 → 2 × 10   40 → 2 × 20   60 → 2 × 30   80 → 2 × 40   100 → 2 × 50

30 → 3 × 10   90 → 3 × 30   60 → 3 × 20   50 → 5 × 10   100 → 5 × 20

3. Multiply and find the product. The first factor is three bronze Knuts. The first task is done for

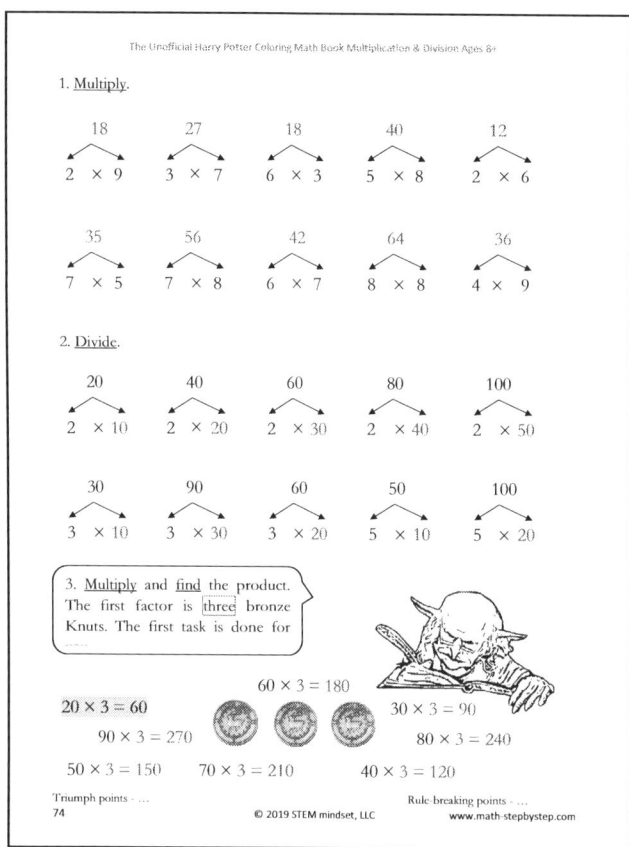

20 × 3 = 60   60 × 3 = 180   30 × 3 = 90
90 × 3 = 270   80 × 3 = 240
50 × 3 = 150   70 × 3 = 210   40 × 3 = 120

Triumph points - …   Rule-breaking points - …
74   © 2019 STEM mindset, LLC   www.math-stepbystep.com

Your home is My Dream School!.. Are there any scholarships for Muggles?!

Don't know. Try to solve the problem.

1. There are 32 pixies in all. I divided the pixies into groups of 3. How many pixies are small?

(30 ÷ 3) × 1 + 1 = 11 (small pixies)

Triumph points - …   Rule-breaking points - …
www.math-stepbystep.com   © 2019 STEM mindset, LLC   75

I wonder, how do you divide by 10 or 100 or even 1000?

Is he – a bit mad? (see *Harry Potter and the Sorcerer's Stone* page 101).

I'm not so sure… Nevermind, look, I divide 80 by 20:

Hint! Step 1: Both numbers have 0's, so, cross out these 0's:

Step 2: divide the digits:

Step 3: write the answer:

80 ÷ 20 = …
↓
8̶0̶ ÷ 2̶0̶ = …
↓
8̶0̶ ÷ 2̶0̶ = 8 ÷ 2 = …
↓
8̶0̶ ÷ 2̶0̶ = 8 ÷ 2 = 4

Just keep in mind that you need to cross out the SAME number of 0's in both the dividend and the divisor! Let me show you:

600 ÷ 30 = Cross out only one 0 (zero) = 6̶0̶0 ÷ 3̶0 = 60 ÷ 3 = 20.

… an' don' ask me questions just now, I think I'm gonna be sick (see *Harry Potter and the Sorcerer's Stone* page 65).

Show me another way!

Triumph points - …   Rule-breaking points - …
76   © 2019 STEM mindset, LLC   www.math-stepbystep.com

I hope the Muggles didn't give you a hard time (see Harry Potter and the Prisoner of Azkaban page 9).

I will help you! Look, I divide 160 by 20.

$160 \div 20 = \ldots$

How many 20's are in 160?
The closest answer is eight 20's:
Check your answer:
Step 1: $8 \times 0 = 0$
Step 2: $8 \times 2 = 16$
Step 3: $160 - 160 = 0$

$$\begin{array}{r} 8 \\ 20\overline{)160} \end{array} \text{ Step1}$$
8×0

$$\begin{array}{r} 8 \\ 20\overline{)160} \\ -160 \\ \hline 0 \end{array} \text{ Step2}$$
8×2

And you can always multiply to check your answer:
$8 \times 20 = 8 \times 2 \times 10 = 16 \times 10 = 160.$

1. Divide. Cross out 0's in both the dividend and the divisor.

150 ÷ 10 = 15   300 ÷ 30 = 30   600 ÷ 20 = 30

840 ÷ 10 = 84   80 ÷ 40 = 2   90 ÷ 30 = 3

550 ÷ 10 = 55   120 ÷ 30 = 4   140 ÷ 70 = 2

560 ÷ 80 = 7   810 ÷ 90 = 9   360 ÷ 40 = 9

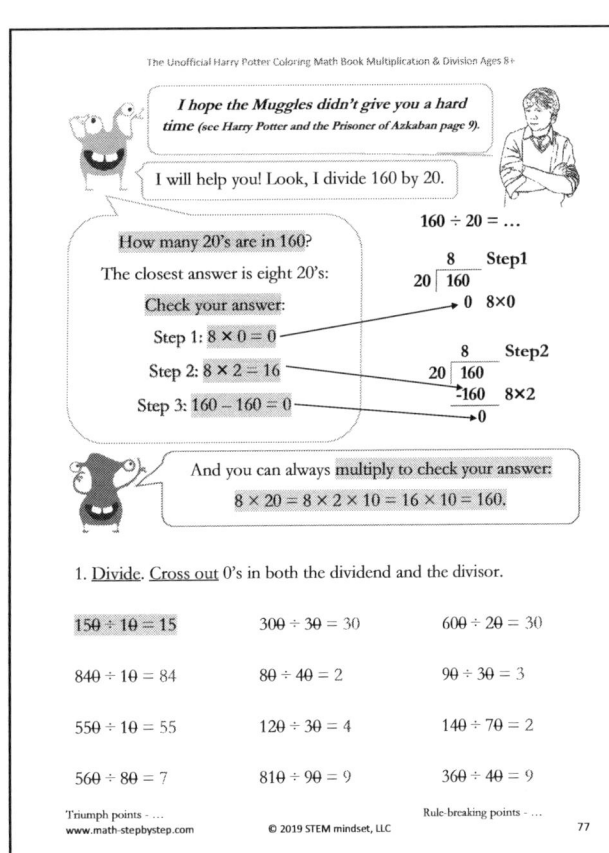

1. Answer the question.

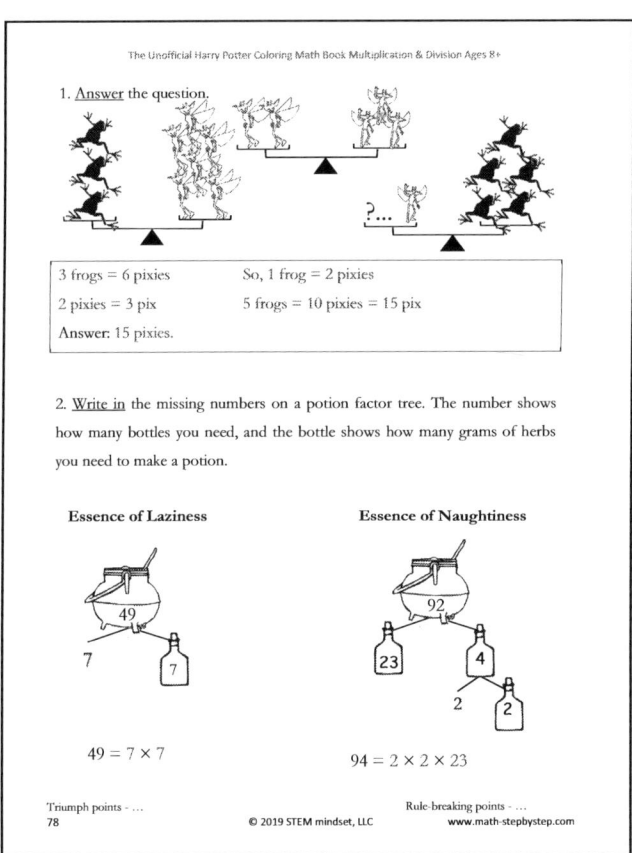

3 frogs = 6 pixies   So, 1 frog = 2 pixies
2 pixies = 3 pix   5 frogs = 10 pixies = 15 pix
Answer: 15 pixies.

2. Write in the missing numbers on a potion factor tree. The number shows how many bottles you need, and the bottle shows how many grams of herbs you need to make a potion.

Essence of Laziness

49
/ \
7   7

$49 = 7 \times 7$

Essence of Naughtiness

92
/ \
23 4
   / \
   2 2

$94 = 2 \times 2 \times 23$

1. Multiply.

24   42   72   35   9
/\   /\   /\   /\   /\
4 × 6   7 × 6   8 × 9   5 × 7   3 × 3

21   30   12   32   45
/\   /\   /\   /\   /\
3 × 7   5 × 6   4 × 3   8 × 4   9 × 5

2. Divide. Cross out 0's.

20   40   60   80   100
/\   /\   /\   /\   /\
20 × 1   20 × 2   20 × 3   20 × 4   20 × 5

30   90   60   80   100
/\   /\   /\   /\   /\
30 × 1   30 × 3   30 × 2   40 × 2   50 × 2

3. Divide and find the quotient. The divisor is three bronze Knuts. The first task is done for you.

180 ÷ 3 = 90   90 ÷ 3 = 30   120 ÷ 3 = 40
270 ÷ 3 = 90   210 ÷ 3 = 70
60 ÷ 3 = 20   240 ÷ 3 = 80   150 ÷ 3 = 50

1. Find and circle or cross out the words to find out more Hogwarts' Professors.

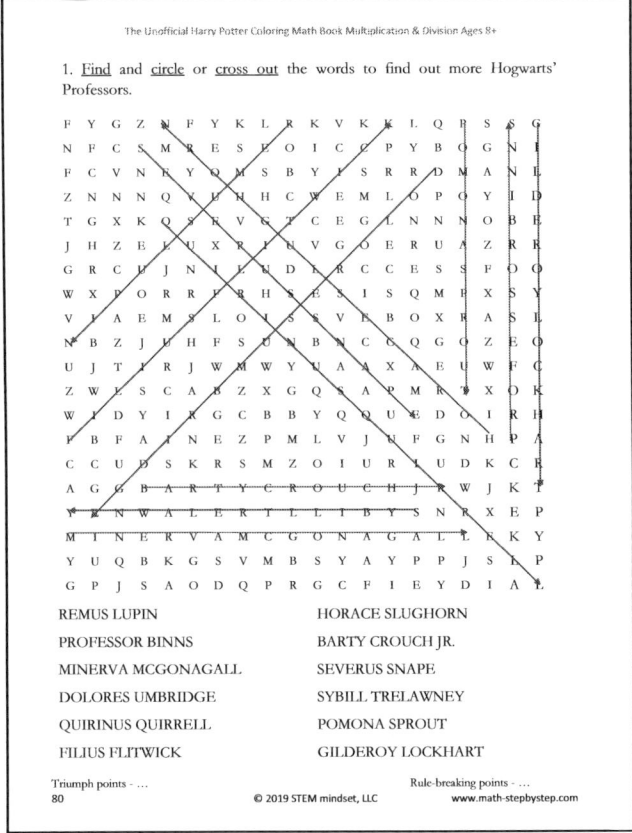

REMUS LUPIN   HORACE SLUGHORN
PROFESSOR BINNS   BARTY CROUCH JR.
MINERVA MCGONAGALL   SEVERUS SNAPE
DOLORES UMBRIDGE   SYBILL TRELAWNEY
QUIRINUS QUIRRELL   POMONA SPROUT
FILIUS FLITWICK   GILDEROY LOCKHART

Page 81

1. Multiply.

```
   70      90      40      70      50
 ×  5    ×  8    ×  2    ×  4    ×  9
 ---     ---     ---     ---     ---
  350     720      80     280     450

   30      70      80      90      60
 ×  4    ×  7    ×  3    ×  7    ×  6
 ---     ---     ---     ---     ---
  120     490     240     630     360

   40      70      80      30      20
 ×  5    ×  8    ×  5    ×  6    ×  8
 ---     ---     ---     ---     ---
  200     560     400     180     160
```

2. Mr. Dursley paid $24 in all for 6 tickets to the zoo. How many three-dollar tickets did he buy? How many five-dollar tickets did he buy?

Paid in all	Tickets in all	Three-dollar tickets	Five-dollar tickets
24	6	? 3	? 3

3-dollar ticket = x; 5-dollar ticket = y
$x + y = 6$ $x = 6 - y$
$3 \times x + 5 \times y = 24$ $3(6-y)+5y=24$
$18-3y+5y=24$ → $2y=6$
$y = 3$ $x = 3$

Answer: 3 tickets; 3 tickets.

Page 82

1. Divide.

```
     90          70          60          70
  2)180       3)210       4)240       5)350
    18          21          24          35
    --          --          --          --
     0           0           0           0

     70          80          70          50
  8)560       5)400       7)490       6)300
    56          40          49          30
    --          --          --          --
     0           0           0           0

     70          80          90          20
  7)490       8)640       9)810       8)160
    49          64          81          16
    --          --          --          --
     0           0           0           0
```

2. Write the missing one-digit numbers where the remainder is the greatest.

$41 \div 7 = 5\ r\ 6$ $41 \div 6 = 6\ r\ 5$ $26 \div 3 = 8\ r\ 2$

$39 \div 4 = 9\ r\ 3$ $13 \div 2 = 6\ r\ 1$ $47 \div 8 = 5\ r\ 7$

$71 \div 9 = 7\ r\ 8$ $53 \div 6 = 8\ r\ 5$ $24 \div 5 = 4\ r\ 4$

Page 83

1. Write in the missing numbers on a potion factor tree. The number shows how many bottles you need, and the bottle shows how many grams of herbs you need to make a potion.

Wideeye Potion

74 — 2, 37

$74 = 2 \times 37$

Beauty-making Potion Antidote

80 — 8, 10; 8 — 4, 2; 4 — 2, 2; 10 — 2, 5

$80 = 2 \times 2 \times 2 \times 2 \times 5$

Beauty-making Potion

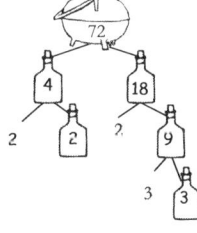

72 — 4, 18; 4 — 2, 2; 18 — 9, 2; 9 — 3, 3

$72 = 2 \times 2 \times 2 \times 3 \times 3$

Burning Potion

76 — 2, 38; 38 — 2, 19

$76 = 2 \times 2 \times 19$

Page 84

1. Answer the questions.

Continue a series of numbers (multiplication).

4, 8, 24, 48, 48×3=144, 144×2 = 288

Think, Brainer, think!

Don't let the Muggles get you down! (see *Harry Potter and the Prisoner of Azkaban* page 10).

2. Find the .

 ÷ (36 − 27) = 9   (34 + 29) ÷  = 7

X = 9 × 9   X = 63 ÷ 7

 = 81    = 9

3. Multiply and find the product. The first factor is five bronze Knuts.

$60 \times 5 = 300$ $20 \times 5 = 100$

$90 \times 5 = 450$

$50 \times 5 = 250$ $70 \times 5 = 350$ $30 \times 5 = 150$

$80 \times 5 = 400$

$40 \times 5 = 200$

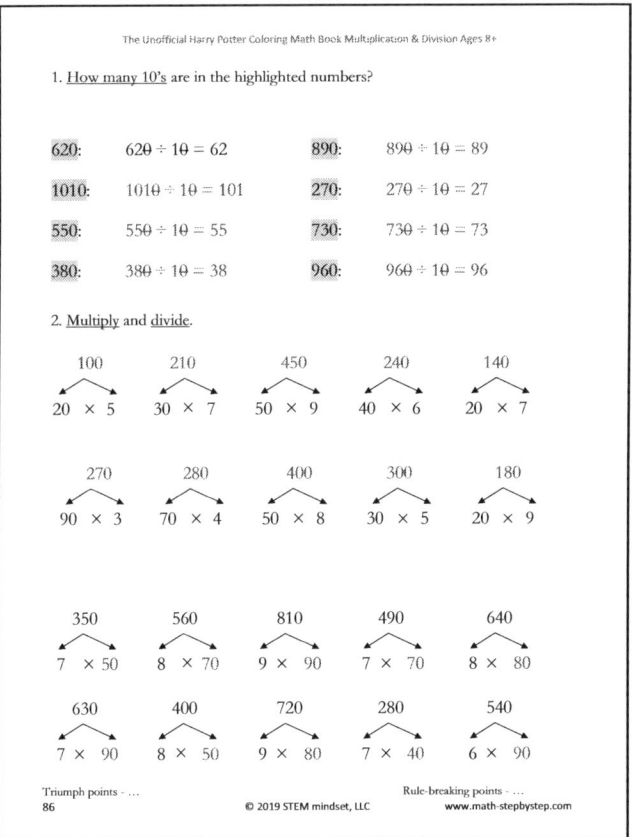

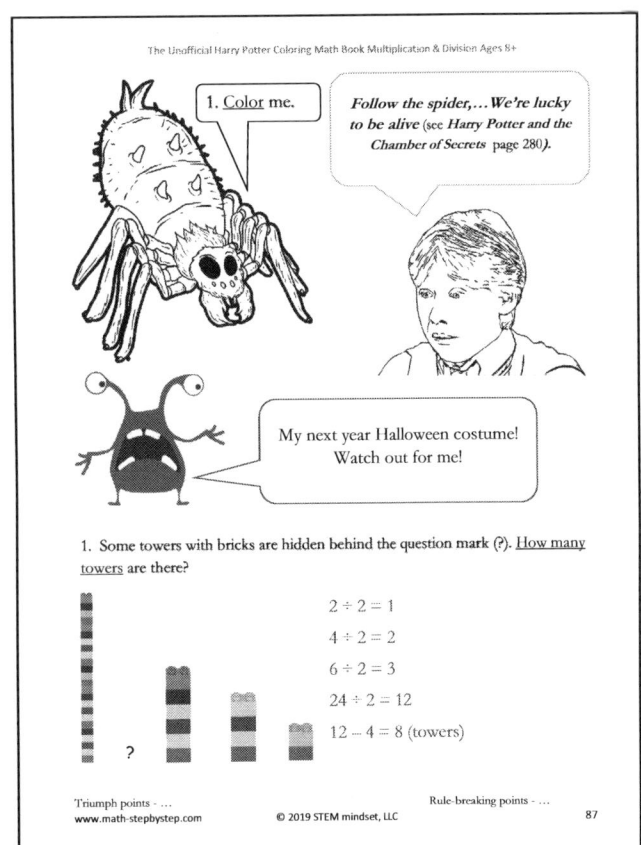

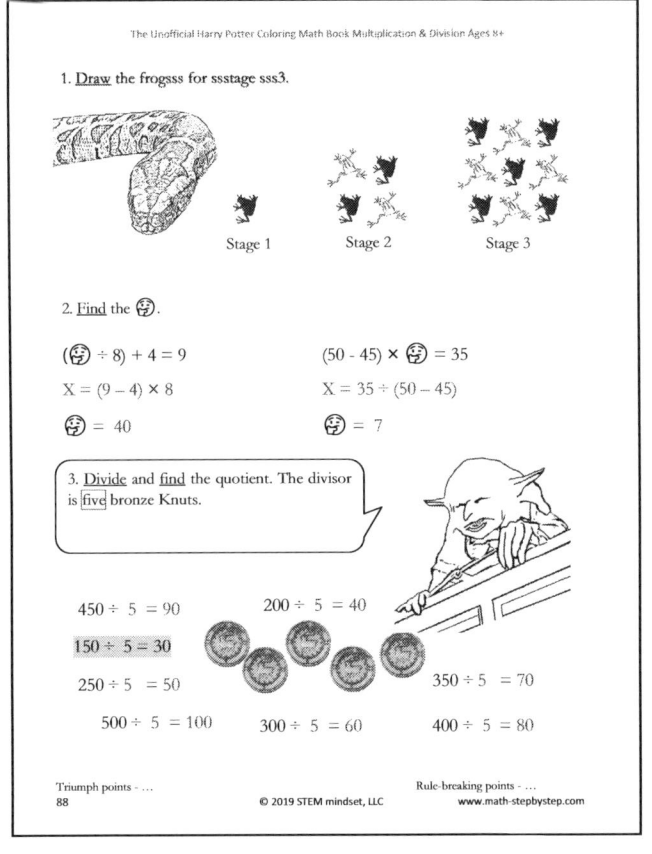

1. <u>Answer</u> the question.

   🧚 × 🧚 = 4800   🧚 = ? 60

   🧚 × 🧚 = 4000   🧚 = ? 80

   🧚 × 🧚 = 3000   🧚 = ? 50

2. There are $\boxed{13}$ flying cars and flying bikes in Hogwarts' parking lot. There are a total of $\boxed{40}$ wheels. <u>How many flying cars</u> are in the Hogwarts' parking lot? <u>How many flying bikes</u> are in the parking lot?

Cars & bikes	Wheels in all	Cars	Bikes
13	40	? 7	? 6

$7 \times 4 = 28$

$6 \times 2 = 12$

Answer: 7 cars and 6 bikes.

---

1. <u>Answer</u> the questions.

1. <u>How many 20's</u> are in the highlighted numbers?

   800:   ̶8̶0̶0̶ ÷ ̶2̶0̶ = 40   680:   ̶6̶8̶0̶ ÷ ̶2̶0̶ = 34

   1000:   ̶1̶0̶0̶0̶ ÷ ̶2̶0̶ = 50   440:   ̶4̶4̶0̶ ÷ ̶2̶0̶ = 22

<u>How many 40's</u> are in the highlighted numbers?

   800:   ̶8̶0̶0̶ ÷ ̶4̶0̶ = 20   480:   ̶4̶8̶0̶ ÷ ̶4̶0̶ = 12

<u>How many 50's</u> are in the highlighted numbers?

   550:   ̶5̶5̶0̶ ÷ ̶5̶0̶ = 11   1500:   ̶1̶5̶0̶0̶ ÷ ̶5̶0̶ = 30

2. <u>Multiply</u> <u>and</u> find the product. The first factor is $\boxed{\text{four}}$ bronze Knuts.

$50 \times 4 = 200$   $90 \times 4 = 360$

$40 \times 4 = 160$ $60 \times 4 = 240$

      $80 \times 4 = 320$

$20 \times 4 = 80$ $70 \times 4 = 280$ $30 \times 4 = 120$

3. <u>Divide</u> and find the quotient. The divisor is four bronze Knuts.

 *Training for the ballet ... ?* (see Harry Potter and the Chamber of Secrets page 171). 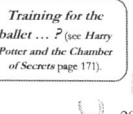  $80 \div 4 = 20$

          $200 \div 4 = 50$ $360 \div 4 = 90$

          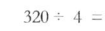 $320 \div 4 = 80$

          $280 \div 4 = 70$

$240 \div 4 = 60$ $160 \div 4 = 40$ $120 \div 4 = 30$

---

1. <u>Multiply</u>. <u>Use</u> five multiplication ways for each problem (decompose the factors). <u>Manipulate</u> the factors!

$15 \times 4 = 5 \times 3 \times 2 \times 2 = 15 \times 1 \times 4 \times 1 = 5 \times 3 \times 2 \times 2 = 10 \times 6 =$
$= 30 \times 2 = 60$

$18 \times 6 = 6 \times 3 \times 2 \times 3 = 2 \times 9 \times 2 \times 3 = 2 \times 3 \times 3 \times 2 \times 3 =$
$= 4 \times 27 = 12 \times 9 = 96$

$20 \times 4 = 5 \times 4 \times 2 \times 2 = 5 \times 2 \times 2 \times 2 \times 2 = 5 \times 2 \times 8 = 10 \times 8 =$
$= 1 \times 80 = 80$

$25 \times 4 = 5 \times 5 \times 2 \times 2 = 10 \times 5 \times 2 = 5 \times 5 \times 4 = 50 \times 2 =$
$= 25 \times 2 \times 2 = 100$

 2. <u>Help</u> the Brainer get the Triwizard Cup.

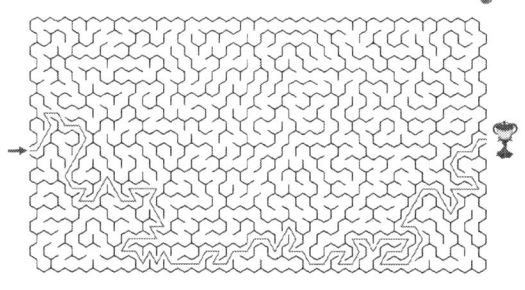

---

1. <u>Evaluate</u> each expression. First, start with the parentheses (brackets), then, division or multiplication OR addition or subtraction from left to right! <u>Indicate</u> the order of operation.

$(54 \div 9)^1 + (18 \div 9)^4 - (18 \div 6)^2 = 5$ $(81 \div 9)^1 \div (54 \div 6)^3 = 1$

$(27 \div 9) + (36 \div 6) - (36 \div 9) = 5$ $(48 \div 6) \div (72 \div 9) = 1$

$(63 \div 9) + (42 \div 6) - (81 \div 9) = 5$ $(30 \div 6) \times (45 \div 9) = 25$

$(45 \div 9) + (72 \div 8) - (27 \div 9) = 11$ $(72 \div 9) \div (32 \div 8) = 2$

$(18 \div 9) + (63 \div 9) - (56 \div 8) = 2$ $(81 \div 9) - (24 \div 8) = 3$

$(16 \div 8) + (36 \div 9) - (45 \div 9) = 1$ $(48 \div 8) \times (54 \div 9) = 36$

$(21 \div 3) + (35 \div 7) - (49 \div 7) = 5$ $(63 \div 9) \times (9 \div 3) = 21$

$(42 \div 7) + (27 \div 3) - (14 \div 7) = 13$ $(28 \div 7) \div (6 \div 3) = 2$

$(21 \div 7) + (24 \div 3) - (35 \div 7) = 6$ $(56 \div 7) \times (12 \div 3) = 32$

$(28 \div 7) + (63 \div 7) - (36 \div 4) = 4$ $(36 \div 4) \div (63 \div 7) = 1$

$(35 \div 7) + (16 \div 4) - (21 \div 7) = 6$ $(56 \div 7) \times (24 \div 4) = 48$

$(49 \div 7) + (20 \div 4) - (42 \div 7) = 6$ $(28 \div 7) \times (8 \div 4) = 8$

Page 93

1. <u>Find</u> and <u>circle</u> or <u>cross out</u> the words to find out more about Severus Snape.

```
K J G Z Z D O A P I Q O R D P E I
D R Q U A G U A S N P G E F O C L
I P O T I O N S M A S T E R T N H
R N P W O D S P N W A R N J E M
N G S O R S A Z X C B O E Q I L
S K B U L E J E I P R F F M X C L
X L F X Y P T H D Y F J D M S A
K G M K T S A L R X R R P A E T
M D W S F I E U P Z E H O K L I
O L U V H P A R Y S A D F H I T W
S D T P N C O C A D U P N B N B O
F P O S V R K H I B O O K U G N
K S V R D D Z N G L I N I L D S K
K S S Y W Q G J V U N E E D N V D
H O G N I V A W D N A W O R E R Q
Y C I N V A D E R Z A C R D G T R
P A Y T C M Z U T K W H W I L W L
```

MIND-READING INVADER
INSUFFERABLE KNOW-IT-ALL
SUBTLE SCIENCE POTION-MAKING
WAND-WAVING CAULDRON
DUNDERHEAD SOPHISTICATED
TEDIOUS PAPERWORK POTIONS MASTER

Page 94

1. <u>Multiply</u>.

```
  90      40      50      40      70
×  5    ×  8    ×  2    ×  4    ×  9
 450     320     100     160     630

  50      60      40      20      80
×  4    ×  7    ×  3    ×  7    ×  6
 200     420     120     140     480

  30      60      50      40      30
×  5    ×  8    ×  5    ×  6    ×  8
 150     480     250     240     240
```

2. <u>Answer</u> the questions.

How many multiples of 7 are between 6 and 66? 9
How many multiples of 4 are between 2 and 33? 8
How many multiples of 9 are between 10 and 85? 8
How many multiples of 6 are between 4 and 59? 9
How many multiples of 3 are between 1 and 31? 10
How many multiples of 8 are between 5 and 52? 6

Page 95

1. <u>Divide</u>.

```
       50          50          70          20
   2│100       3│150       4│280       5│100
     10          15          28          10
      0           0           0           0

       80          60          30          70
   8│640       5│300       7│210       6│420
     64          30          21          42
      0           0           0           0

       90          90          60          20
   7│630       8│720       9│540       8│160
     63          72          54          16
      0           0           0           0
```

2. Four cages contained <u>48, 67, 59, and 26</u> pixies respectively. <u>How could Professor Lockhart rearrange</u> the pixies so that each cage contained the <u>same</u> number of pixies?

48+67+59+26 = 200
200÷4 = 50

Answer: Add 2; Subtract 17; Subtract 9; Add 24.

Page 96

1. <u>Write in</u> the missing numbers on a potion factor tree. The number shows how many bottles you need, and the bottle shows how many grams of herbs you need to make a potion.

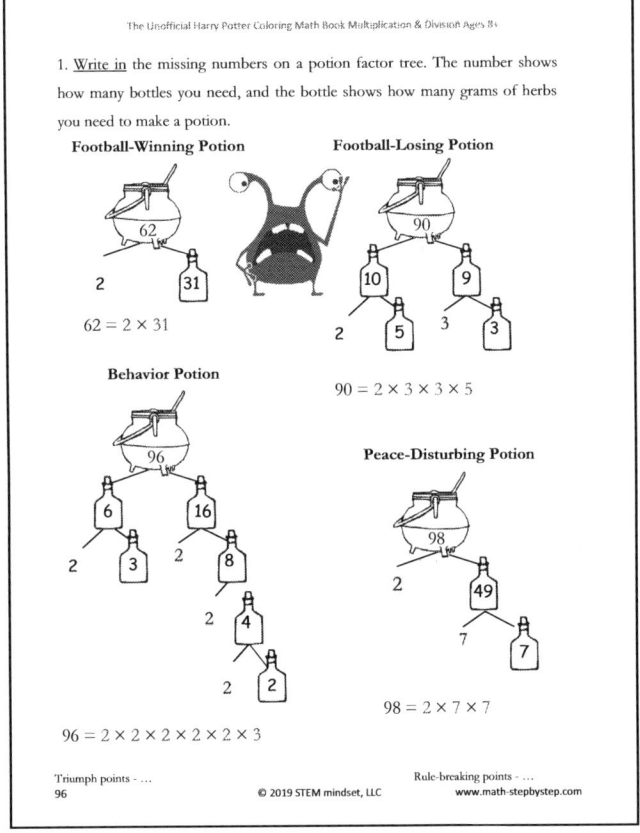

Football-Winning Potion
62 = 2 × 31

Football-Losing Potion
90 = 2 × 3 × 3 × 5

Behavior Potion
96 = 2 × 2 × 2 × 2 × 2 × 3

Peace-Disturbing Potion
98 = 2 × 7 × 7

Page 97

1. Answer the questions. Write one addition and multiplication number sentence for each problem. Indicate the order of operations.

Say you're ill... Really break your leg (see *Harry Potter and the Sorcerer's Stone* page 176).

How many black circles are there?

(5×8)+(2×3)+(2×4)+2= = 56

How many black triangles are there?

6 × 6 = 36

2. Find the ☺.

Just do your best, (see *Harry Potter and the Sorcerer's Stone*

(87 + 63) ÷ ☺ = 30
X = 150 ÷ 30
☺ = 5

(80 + ☺) ÷ 4 = 30
X = (30 × 4) − 80
☺ = 40

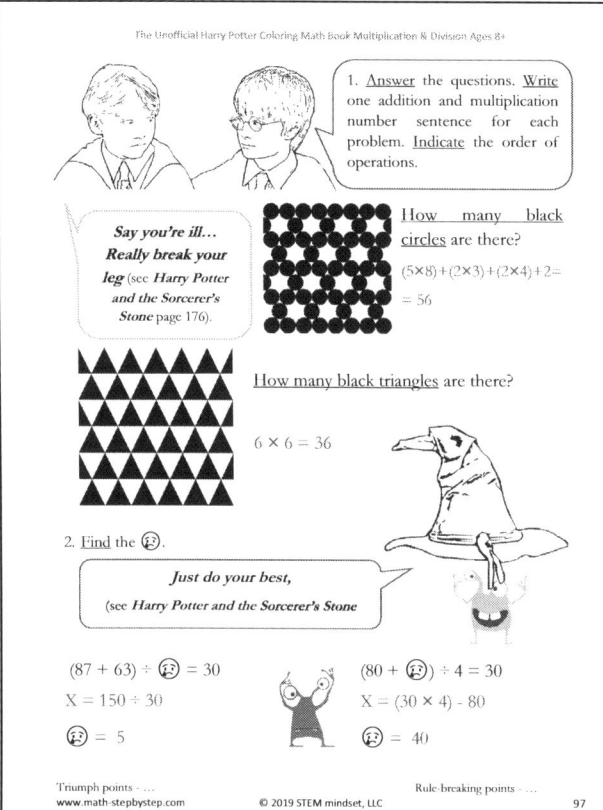

Page 98

1. Answer the questions.

1. How many 30's are in:

240: 240 ÷ 30 = 8   180: 180 ÷ 30 = 6
360: 360 ÷ 30 = 12   120: 120 ÷ 30 = 4
270: 270 ÷ 30 = 9   210: 210 ÷ 30 = 7

Learn math and be around to save the day! Bad Dobby!

How many 40's are in the highlighted numbers?

240: 240 ÷ 40 = 6   320: 320 ÷ 40 = 8
360: 360 ÷ 40 = 9   280: 280 ÷ 40 = 7
200: 200 ÷ 40 = 5   160: 160 ÷ 40 = 4

2. Multiply and find the product. The first factor is six bronze Knuts.

90 × 6 = 540   60 × 6 = 360
50 × 6 = 300   80 × 6 = 480
40 × 6 = 240
20 × 6 = 120   70 × 6 = 420   30 × 6 = 180

3. Divide and find the quotient. The divisor is six bronze Knuts.

180 ÷ 6 = 30   480 ÷ 6 = 80
420 ÷ 6 = 70   300 ÷ 6 = 50
540 ÷ 6 = 90   360 ÷ 6 = 60
240 ÷ 6 = 40   120 ÷ 6 = 20

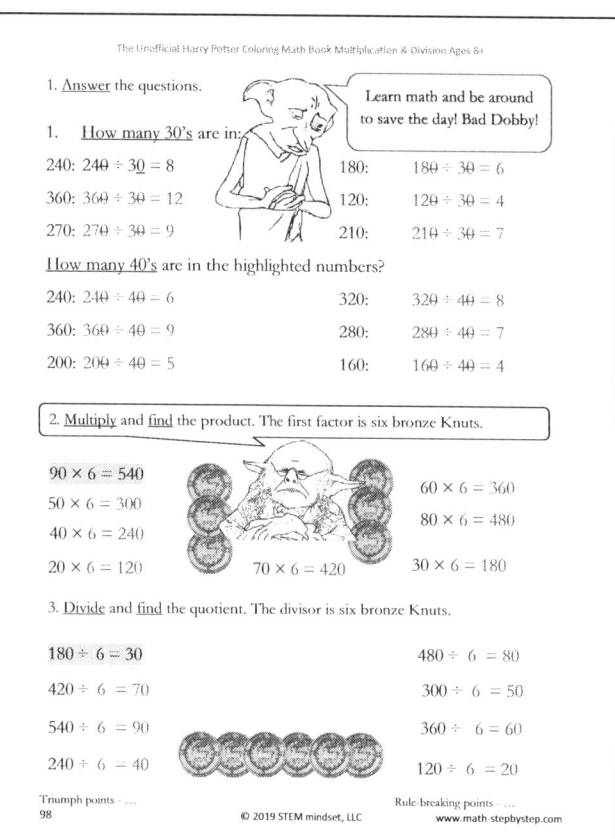

Page 99

1. Find and circle or cross out the words to find out more about Ron Weasley.

```
E X H R D P V I M X B M V Q U P P
R U D E A N T H O M A S J W G I I
J O M Y U Q H M N P C G Y M O G V
R E P E E K H C T I D D I U Q W R
G Q W Z H U L Y K R M F T K I A A
P J V C U A M P Z E S M V Y J D R
S R E B B A C S S Z O T Z F R G T
O G G R I M M A U L D P L A C E H
D C A N L B U T L X Z R M J R O U
W L D M M C Y Y H O W L E R O N R
T M F X R Q W E V Z B G R F H J W
Q Z I S S E H C D R A Z I W Z B E
W I N G A R D I U M L A V I O S A
G Y M S Y R R D J J A B H A B A S
G U V Y B F S D D O K D V K G Q L
R E I L K L A S A V L W K I F A E
Y V J A U F U R R I G Y A U J C Y
```

MOLLY WEASLEY   ARTHUR WEASLEY
CHASER   DEAN THOMAS
GRIMMAULD PLACE   HORCRUX
WIZARD CHESS   HOWLER
QUIDDITCH KEEPER   SCABBERS
WINGARDIUM LAVIOSA   PIGWIDGEON

Page 100

1. Color me.

Caution! My nightmare is on the rampage!

Oy, pea-brain! (see *Harry Potter and the Sorcerer's Stone* page 143)

Oh, no (see *Harry Potter and the Sorcerer's Stone* page 105) but Math first, so, Place the parentheses to make the expressions true.

(5 × 9) ÷ 3 = 15   (50 ÷ 2) × 3 = 75

(18 + 6) ÷ 3 = 8   (3 × 10) ÷ (15 ÷ 3) = 6

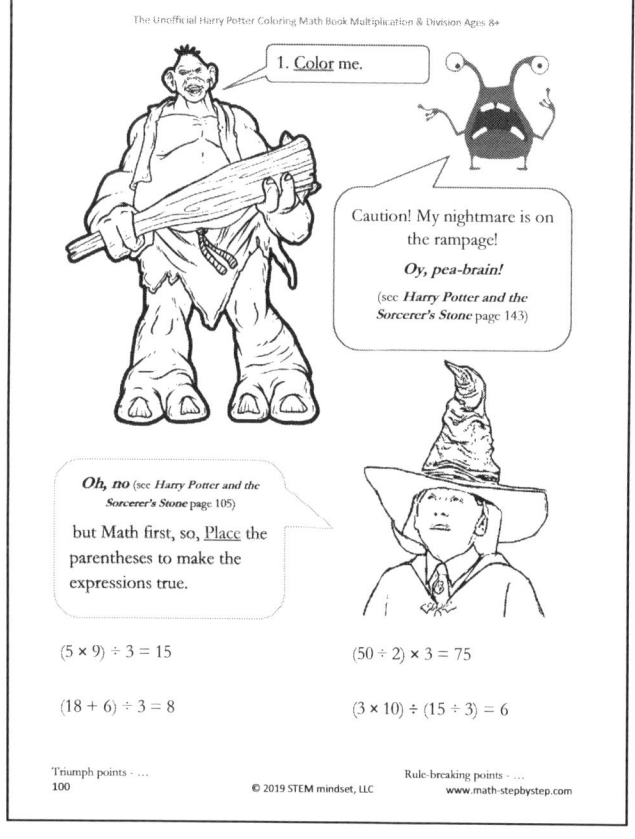
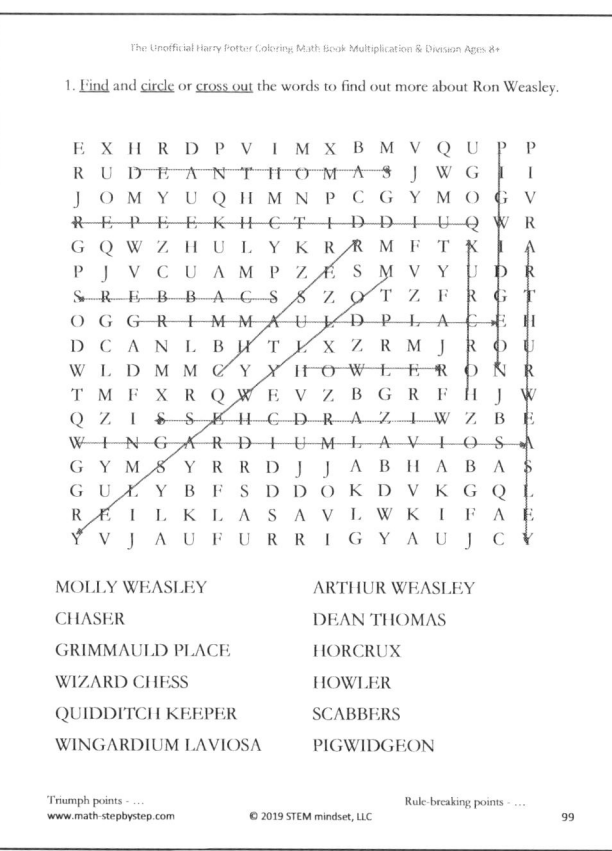

Page 101

1. Anssswer the questionsss.

1. How many 50's are in the highlighted numbers?

350: 350 ÷ 50 = 7   300: 300 ÷ 50 = 6
450: 450 ÷ 50 = 9   250: 250 ÷ 50 = 5
200: 200 ÷ 50 = 4   400: 400 ÷ 50 = 8

How many 60's are in the highlighted numbers?

600: 600 ÷ 60 = 10   300: 300 ÷ 60 = 5
480: 480 ÷ 60 = 8   240: 240 ÷ 60 = 4
540: 540 ÷ 60 = 9   420: 420 ÷ 60 = 7

2. Multiply and find the product. The first factor is seven bronze Knuts.

90 × 7 = 630
50 × 7 = 350   60 × 7 = 420
40 × 7 = 280   80 × 7 = 560
20 × 7 = 140   70 × 7 = 490   30 × 7 = 210

3. Divide and find the quotient. The divisor is seven bronze Knuts.

140 ÷ 7 = 20
490 ÷ 7 = 70   350 ÷ 7 = 50
420 ÷ 7 = 60   210 ÷ 7 = 30
280 ÷ 7 = 40   630 ÷ 7 = 90   560 ÷ 7 = 80

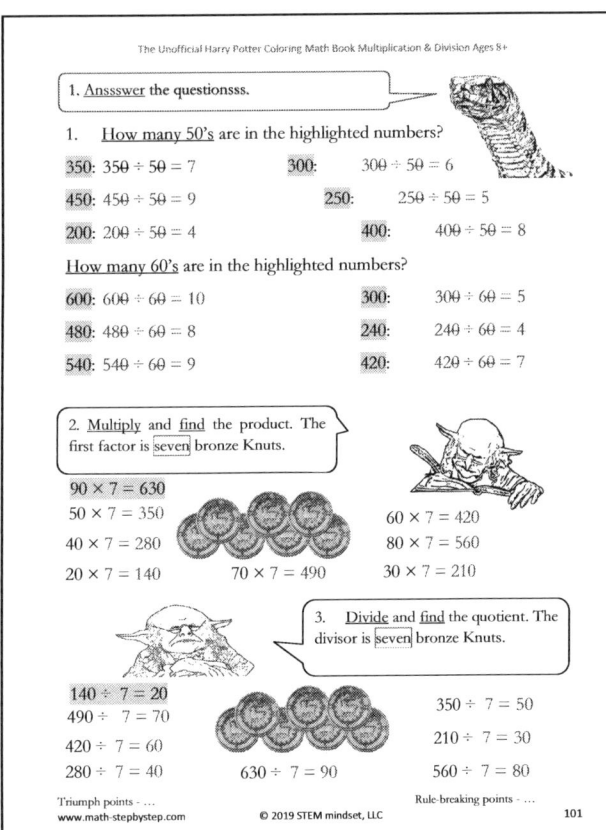

Page 102

1. Circle the right answer.

A 90 × 2 × 2
B 6 × 45

A is greater than B
(A is greater than B)
A is equal to B
A is less than B

A (20 × 40) - 250
B (40 × 20) + 250

A is greater than B
A is equal to B
(A is less than B)

A (80 × 7) + 325
B (20 × 40) + 85

A is greater than B
(A is equal to B)
A is less than B

A 4 × 8 × 2 × 5
B 4 × 4 × 4 × 6

A is greater than B
(A is less than B)
A is equal to B

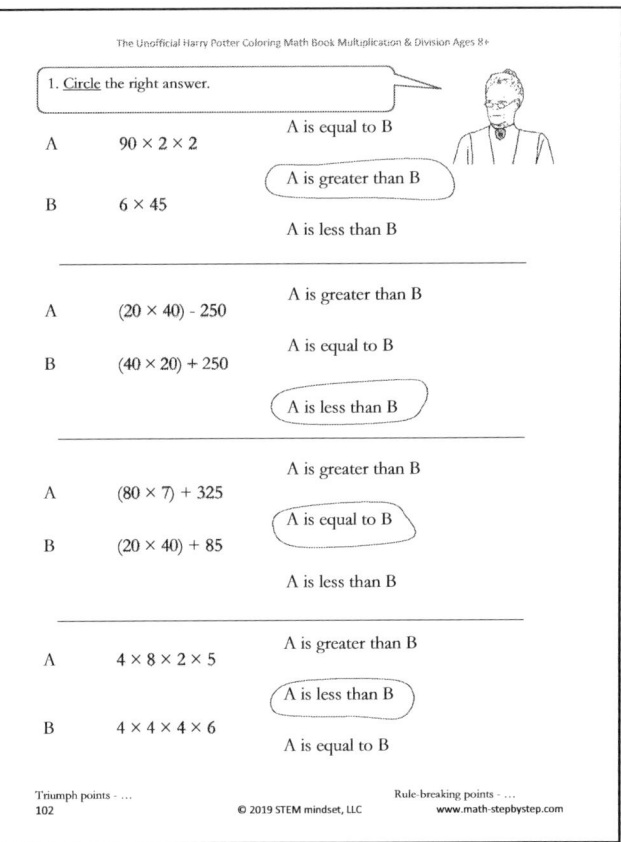

Page 103

1. Answer the questions.

1. How many 70's are in:

700: 700 ÷ 70 = 10   350: 350 ÷ 70 = 5
490: 490 ÷ 70 = 7   280: 280 ÷ 70 = 4
420: 420 ÷ 70 = 6   210: 210 ÷ 70 = 3

How many 80's are in:

800: 800 ÷ 80 = 10   400: 400 ÷ 80 = 5
720: 720 ÷ 80 = 9   560: 560 ÷ 80 = 7
640: 640 ÷ 80 = 8   320: 320 ÷ 80 = 4

2. Multiply and find the product. The first factor is eight bronze Knuts.

50 × 8 = 400   90 × 8 = 720   60 × 8 = 480
40 × 8 = 320   80 × 8 = 640
20 × 8 = 160   70 × 8 = 560   30 × 8 = 240

3. Divide and find the quotient. The divisor is eight bronze Knuts.

160 ÷ 8 = 20   480 ÷ 8 = 60
400 ÷ 8 = 50
640 ÷ 8 = 80   720 ÷ 8 = 90
320 ÷ 8 = 40   240 ÷ 8 = 30   560 ÷ 8 = 70

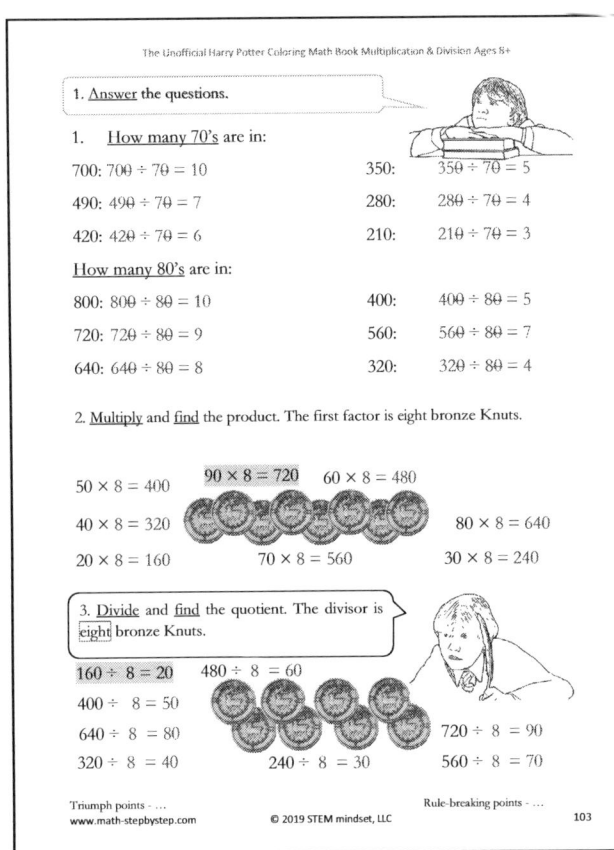

Page 104

1. Multiply. Use five multiplication ways for each problem (decompose the factors).

24 × 4 = 6 × 4 × 2 × 2 = 2 × 3 × 2 × 2 × 2 = 12 × 2 × 2 × 2 =
= 16 × 6 = 4 × 4 × 2 × 3 = 96

30 × 6 = 6 × 5 × 2 × 3 = 2 × 3 × 5 × 2 × 3 = 9 × 4 × 5 = 20 × 9 =
= 15 × 4 × 3 = 180

45 × 8 = 5 × 9 × 2 × 4 = 5 × 3 × 3 × 2 × 2 × 2 = 5 × 6 × 6 × 2 =
= 10 × 36 = 12 × 2 × 15 = 360

50 × 4 = 2 × 25 × 2 × 2 = 2 × 5 × 5 × 2 × 2 = 4 × 10 × 5 = 25 × 8 =
= 40 × 5 = 200

2. Help me get the Triwizard Cup. Broomstick, no! Down!

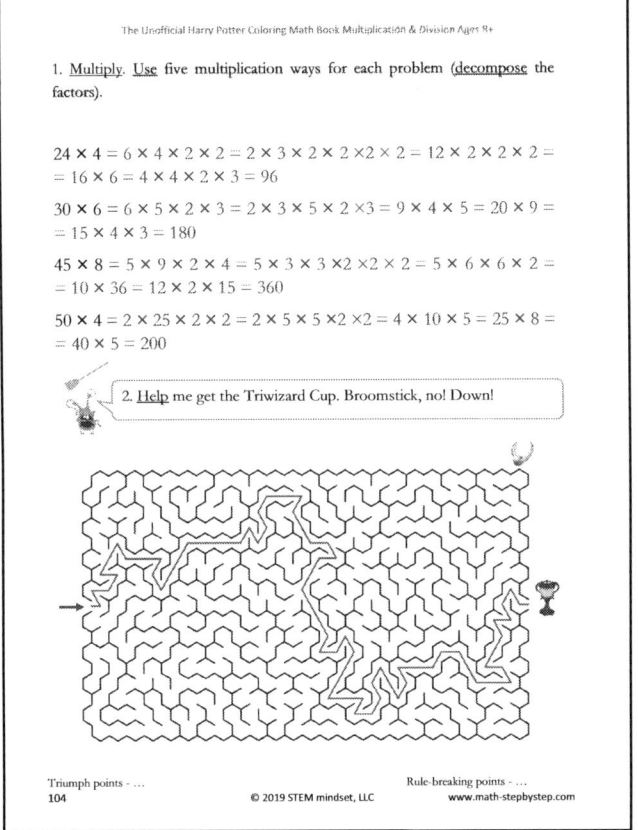

Page 105

1. How many 90's are in the highlighted numbers?

900: 900 ÷ 90 = 10   450: 450 ÷ 90 = 5
810: 810 ÷ 90 = 9   540: 540 ÷ 90 = 6
630: 630 ÷ 90 = 7   720: 720 ÷ 90 = 8

2. Multiply and find the product. The first factor is nine bronze Knuts.

60 × 9 = 540   90 × 9 = 810   80 × 9 = 720
50 × 9 = 450
40 × 9 = 360
20 × 9 = 180   70 × 9 = 630   30 × 9 = 270

3. Divide and find the quotient. The divisor is nine bronze Knuts.

                        180 ÷ 9 = 20
450 ÷ 9 = 50   720 ÷ 9 = 80
540 ÷ 9 = 60   630 ÷ 9 = 70
        270 ÷ 9 = 30   360 ÷ 9 = 40   810 ÷ 9 = 90

4. Find the ☺.

(120 + 240) ÷ ☺ = 90   (119 + ☺) ÷ 5 = 70
X = (120+240) ÷ 90   X = (5 × 70) − 119
☺ = 4   ☺ = 231

Page 106

1. Find and circle or cross out the words to find out more about Professor McGonagall.

G	X	L	I	G	X	K	G	Q	U	T	M	H	A	B	U	D	H
U	L	A	N	Q	N	H	U	T	E	W	Z	N	Y	J	O	L	T
I	L	C	E	K	A	U	Y	M	I	V	L	A	D	J	L	R	Z
N	E	K	P	E	P	R	B	N	F	M	C	H	M	O	A	F	W
E	P	O	T	T	K	K	J	T	A	R	E	Q	R	N	Y	F	Z
A	S	F	I	M	E	T	O	G	N	G	V	T	S	G	U	L	V
P	G	C	T	Y	E	G	U	V	D	G	N	F	U	S	N	L	G
I	N	O	U	P	T	S	V	T	N	I	I	T	X	R	Q	R	I
G	N	D	G	B	I	V	H	A	G	F	T	P	J	N	X	S
S	H	F	E	T	X	Q	Y	T	U	Y	T	P	Z	E	K	E	L
L	C	I	S	A	A	T	N	R	L	L	A	B	E	L	U	Y	R
X	T	D	O	M	G	U	A	D	X	M	T	N	G	F	K	G	H
P	I	E	R	T	O	T	U	M	L	O	C	O	M	O	T	O	R
V	W	N	P	M	I	N	E	E	D	L	E	W	O	R	K	R	D
Z	S	C	H	O	P	W	P	K	P	A	Z	V	T	P	F	E	Q
G	R	E	X	L	L	E	P	S	G	N	I	H	S	I	N	A	V
G	O	K	H	A	V	E	A	B	I	S	C	U	I	T	C	P	R

VANISHING SPELL   TRANSFIGURATION
ANIMAGUS   NEEDLEWORK
TIME-TURNER   PIERTOTUM LOCOMOTOR
SWITCHING SPELL   GUINEA-PIGS
YULE BALL   TIGHT BUN
INEPTITUDES   HAVE A BISCUIT

Page 107

1. Multiply.

```
   40       50       60       90       90
 ×  5     ×  8     ×  2     ×  4     ×  9
  200      400      210      360      810

   60       50       50       60       30
 ×  4     ×  7     ×  3     ×  7     ×  6
  240      350      150      420      180

   80       30       90       40       80
 ×  5     ×  8     ×  5     ×  6     ×  8
  400      240      450      240      640
```

2. Harry counted 9 kittens (k) and owls (o) outside. The kittens and birds had 28 legs altogether. How many owls did Harry see?

Kittens & Birds	Legs	Kittens	Owls	

k+o=9	4×k+2×o= 28	? 5	? 4	k+o=9 k=9−o
				4×k+2×o=28 4(9−o)+2o=28
				                36−4o+2o=28 →o=8÷2
				                o=4 k=5
				Answer: 4 owls.

Page 108

1. Divide.

```
          1 0 0         9 0          5 0          8 0
      2 │ 2 0 0     3 │ 2 7 0    8 │ 4 0 0    5 │ 4 0 0
          2 0           2 7           4 0          4 0
           0             0             0            0

           6 0          9 0           7 0          5 0
      8 │ 4 8 0     5 │ 4 5 0    7 │ 4 9 0    6 │ 3 0 0
          4 8           4 5           4 9          3 0
           0             0             0            0

           6 0          7 0           6 0          9 0
      7 │ 4 2 0     8 │ 5 6 0    9 │ 5 4 0    8 │ 7 2 0
          4 2           5 6           5 4          7 2
           0             0             0            0
```

2. Ron had 70 Chocolate Frog cards. After he gave Harry 7 cards, Ron had nine times as many cards as Harry. How many cards did Harry have?

70 − 7 = 63 (Ron after he gave Harry 7 cards)

63 ÷ 9 = 7 (Harry since Ron had nine times as many cards as Harry)

Answer: Harry had 7 cards.

1. <u>Write in</u> the missing numbers on a potion factor tree. The number shows how many bottles you need, and the bottle shows how many grams of herbs you need to make a potion.

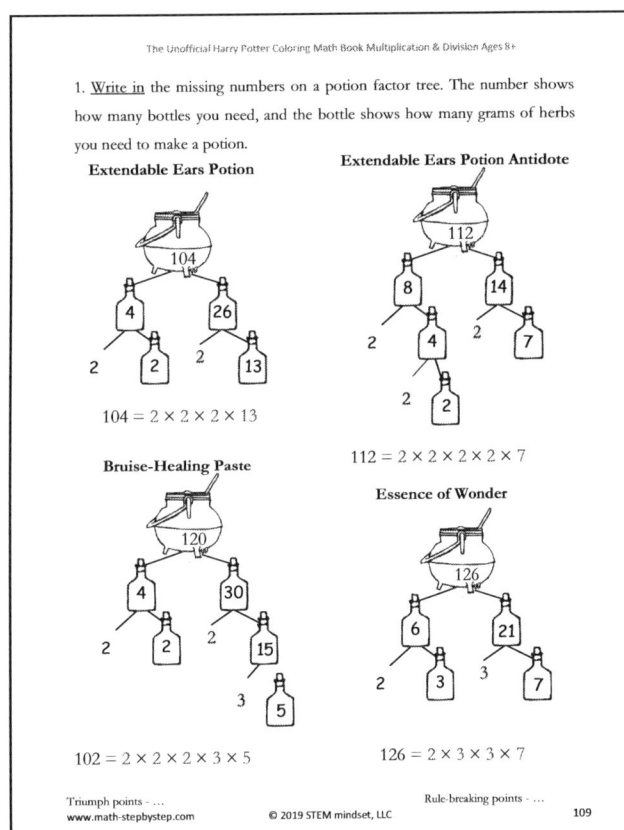

Extendable Ears Potion

$104 = 2 \times 2 \times 2 \times 13$

Extendable Ears Potion Antidote

$112 = 2 \times 2 \times 2 \times 2 \times 7$

Bruise-Healing Paste

$102 = 2 \times 2 \times 2 \times 3 \times 5$

Essence of Wonder

$126 = 2 \times 3 \times 3 \times 7$

1. <u>Place</u> the parentheses to make the expressions true.

$(7 \times 8) \div 4 = 14$ $30 \div (3 \times 2) = 5$

$(10 + 4) \div 7 = 2$ $6 \times (2 \times 5) \div 2 = 30$

Made in the USA
San Bernardino, CA
04 November 2019